Delhi's Changing Built Environment

The rapid expansion, urban form and development of the built environment in the world's second most populous city, Delhi, has been the consequence of social, political, economic, planning and architectural traditions that have shaped the city over thousands of years. Whilst seamless at times, these traditions have often resulted in the fragmented development of the city's built environment. This book charts the political, economic and social forces that drove development in India generally and in Delhi in particular, and investigates the drivers and constituents of Delhi's urban landscape. The book provides a lens through which to examine the development path of a mega-city, which can be used as a guide in the development of emerging urban centres. Furthermore, the strengths and weaknesses of Delhi's built environment are critically analysed, with consideration to the role of the market, finance and policy over time. This book not only provides valuable insight into the physical evolution of Delhi and its surrounds, but it also asks broader questions about how people, power and politics interact with urban environments. It is essential reading for planners, architects, urbanists and social historians.

Piyush Tiwari is Professor of Property at University of Melbourne, Australia. His research interests include infrastructure policy, housing economics and mortgages, commercial real estate investment, and financing infrastructure in developing countries. He has published numerous research papers on issues related to real estate and infrastructure.

Jyoti Rao is an early career researcher pursuing doctoral research at the University of Melbourne. Issues related to housing and land economics are of prime interest to her. She is professionally trained as an architect, urban planner and real estate professional.

Routledge Studies in International Real Estate

The Routledge Studies in International Real Estate series presents a forum for the presentation of academic research into international real estate issues. Books in the series are broad in their conceptual scope and reflect an inter-disciplinary approach to Real Estate as an academic discipline.

Oiling the Urban Economy
Land, Labour, Capital, and the State in Sekondi-Takoradi, Ghana
Franklin Obeng-Odoom

Real Estate, Construction and Economic Development in Emerging Market Economies
Edited by Raymond T. Abdulai, Franklin Obeng-Odoom, Edward Ochieng and Vida Maliene

Econometric Analyses of International Housing Markets
Rita Li and Kwong Wing Chau

Sustainable Communities and Urban Housing
A Comparative European Perspective
Montserrat Pareja Eastaway and Nessa Winston

Regulating Information Asymmetry in the Residential Real Estate Market
The Hong Kong Experience
Devin Lin

Delhi's Changing Built Environment
Piyush Tiwari and Jyoti Rao

Delhi's Changing Built Environment

Piyush Tiwari and Jyoti Rao

LONDON AND NEW YORK

First published 2018
by Routledge
2 Park Square, Milton Park, Abingdon, Oxon OX14 4RN

and by Routledge
605 Third Avenue, New York, NY 10017

First issued in paperback 2021

Routledge is an imprint of the Taylor & Francis Group, an informa business

Publisher's Note
The publisher has gone to great lengths to ensure the quality of this reprint but points out that some imperfections in the original copies may be apparent.

British Library Cataloguing-in-Publication Data
A catalogue record for this book is available from the British Library

Library of Congress Cataloging-in-Publication Data
Names: Tiwari, Piyush, author. | Rao, Jyoti, author.
Title: Delhi's changing built environment / Piyush Tiwari and Jyoti Rao.
Description: Abingdon, Oxon ; New York, NY : Routledge, 2018. | Series: Routledge studies in international real estate | Includes bibliographical references.
Identifiers: LCCN 2017044395 | ISBN 9781138907584 (hardback : alk. paper) | ISBN 9781315695037 (ebook)
Subjects: LCSH: Urbanization—India—Delhi—History. | Human settlements—India—Delhi—History. | City planning—India—Delhi—History. | Architecture—India—Delhi—History. | Landscape changes—India—Delhi—History.
Classification: LCC HT384.I42 D4585 2018 | DDC 307.760954/56—dc23
LC record available at https://lccn.loc.gov/2017044395

ISBN 13: 978-1-03-209585-1 (pbk)
ISBN 13: 978-1-138-90758-4 (hbk)

Typeset in Goudy
by Apex CoVantage, LLC

Ik roz apni rooh se poocha, ke dilli kya hai.
To yun Jawab me keh gayi,
Ye duniya maano jism hai aur dilli uski jaan

– Mirza Ghalib

[One day I asked my soul, what is Delhi and it replied- if world is the body then Delhi is its soul.]

This book is a tribute to all the great people, past and present, who made Delhi and its soul.

Contents

Figures

Tables

1 Delhi and its surrounds

An introduction

Delhi has been an epitome of India's history with its succession of glory and disaster and with its great capacity to absorb many cultures and yet remain itself. It is a gem with many facets, some bright and some darkened by age, presenting the course of India's life and thought during the ages.

—Pandit Jawahar Lal Nehru, 1958

1.1 Introduction

Built environment is a dynamic intervention of changing human needs, wants, thoughts, actions, power, whims and fancies into the natural environment that creates a vibrant mosaic of urban form, materials, architectural design and space. At times these interventions are grand, the impact of which is positive on the quality of life for generations. At other times, human actions are shortsighted and result in sub-optimal spaces for human activities and also negatively impact the environments. Delhi and its surrounds offer a cradle for remarkable civilisations and cultural confluences that have impacted the built environment over many centuries. The richness of the city's built environment is reflected in its continuity through time despite merging of cultures and emergence of new 'Delhi' culture not once, not twice but many times. Sometimes this has left fissures and discontinuities too in the urban fabric of the city, which becomes a disjointed piece. Delhi is said to be a city of eight cities, which took birth, blossomed and perished as the empires that gave birth to them perished. Though the importance of one city declined with the emergence of another, this did not mean that the older city degenerated to brink. Rather the old merged with the new – the young and dynamic. The new inherited the features of the old. While reinventing, many features, forms, shape and materials of the old were carried forward from the old to the new creating a unique style. The visible built environment of modern-day Delhi and its surrounds is a continuum of more than 1,500 years of human actions that is reflected in its buildings, urban form, urban systems and materials. This does not mean that civilisations did not exist before that, or they did not influence the built environment in Delhi and its surrounds. As would be discussed later, civilisations have existed approximately since 1500 BC, but the scope of this book is limited to the discussion of the built environment of the period through which traces of reasonable size and shape exist to this date.

The built environment of a city is influenced by its political, economic and social systems that evolve over a period of time. These form the context within which cities take shape. In this introduction to *Delhi's Changing Built Environment*, we would begin with an introduction to this context within which we will examine the built environment. It may be highlighted here that the book is not a historical treatise of Delhi, but it uses historical events to understand the political, social and economic forces that shaped the land and built space. Before we do that, we need to define what is built environment, and this is dealt with in Section 1.2. This is followed by a discussion on the positioning of the book within a multi-disciplinary context in Section 1.3. Section 1.4 discusses a workable time period to discuss political and social evolution of Delhi and its surrounds. The chronology of eight cities that make up Delhi is discussed. The political, economic and social contexts within which the built environment of Delhi and surrounds evolved are discussed in Section 1.5. Section 1.6 briefly gives an overview of chapters in the book.

1.2 What is built environment?

Bartuska (2007) defines built environment as "everything humanly made, arranged or maintained to fulfil human purposes (needs, wants and values) to mediate the overall environment with results that affect the environmental context". The needs that built environmental fulfils are psychological and social. In addition, the built environment is an expression of personal and collective values. These values are subjective as they deal with beliefs, opinions and attitudes. These attitudes find expression in built environment. As an example, the eight cities of Delhi are a reflection of rulers' attitude towards religion, polity and society. Given that human purposes are manifold and they are dynamic over time, changes that people make to their environment are "extensive expression of past and present cultures" (ibid). The resulting cities are the most complex human systems that are ever created, with numerous dynamic linkages over space between humans and their activities.

Bartuska (2007) identifies seven components of built environment. Products such as materials (bricks and mortar, concrete and steel, wood, polymers and plastics, machines and tools etc.) are the most fundamental component of the built environment. These are used to perform specific tasks. The use and availability of products is as much a function of political, social and economic contexts as the technology. The second component is the interiors. These are the spaces "defined by an arranged grouping or products and generally enclosed within a structure" (ibid). Spaces such as living room, workroom, private room, auditoriums, offices, religious places etc. are created to perform activities and mediate external forces. The third component is structures, "planned groupings of spaces defined by and constructed of products" (ibid). Structures are a combination of related activities. Examples are housing, offices, temples and churches, schools, bridges, tunnels. Landscapes are the fourth component of built environment comprising "exterior areas and/or settings for planned groupings of spaces and structures" (ibid). The fifth component of the built environment is cities, "grouping of structures and landscapes of varying sizes and complexities generally clustered together to define a community for economic, social, cultural and/or environmental reasons" (ibid). Regions, the sixth component of the built environment, are "groupings of cities and landscapes of various sizes and complexities" which are "generally defined by common political, social, economic and/or environmental characteristics" (ibid). The last component of built environment is the earth, encompassing all other components.

To examine the built environment of Delhi and its surrounds, we reclassify six components of built environment discussed earlier (leaving the last component, earth, from discussion as it encompasses everything and is not part of our scope for discussion) in two themes (products, interiors and structures are discussed under the theme "continuities and discontinuities in design", landscape, cities and regions are discussed under the theme "urban form and imageability") to help in forming a workable structure for this book (Figure 1.1).

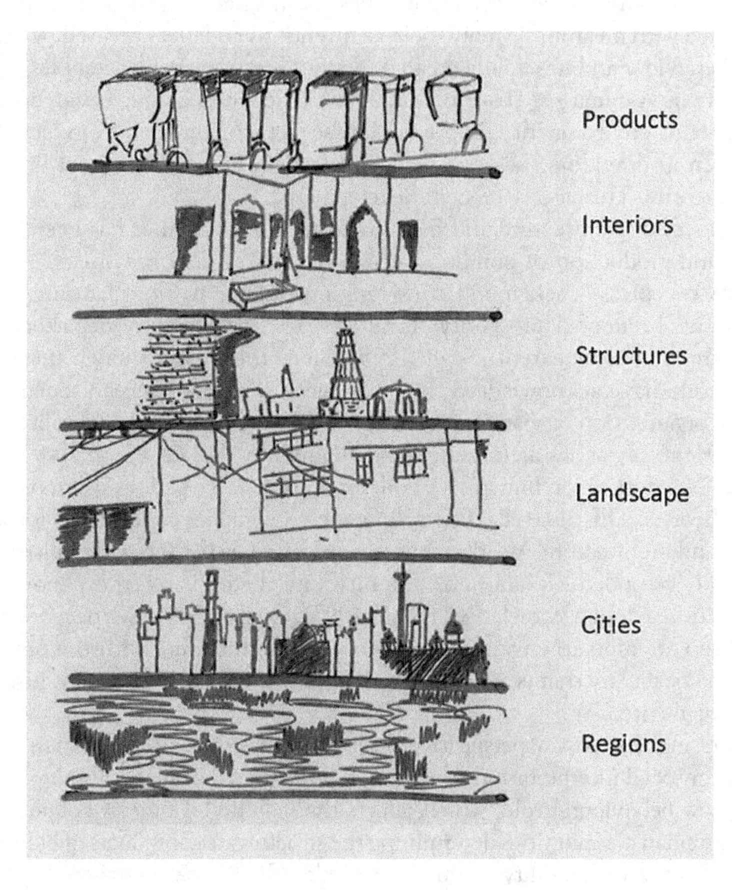

Figure 1.1 Components of a built environment

Source: Authors

1.3 An institutional pyramid

Existing and archaeological evidences pertaining to built environment and its components have been studied by different disciplines to develop theories, but these, merely as a tool, are less than satisfactory utilisations of these materials. Archaeologists, who are concerned with studying human history through excavation of sites and analysis of artifacts, in the absence of their own theoretical strand, have tended to use the materials excavated to explain social life using social theories. As articulated by Harrington (2005), "social theory can be defined as the study of scientific ways of thinking about social life. It encompasses ideas about how societies change and develop, about methods of explaining social behaviour, about power and social structure, class, gender and ethnicity, modernity and 'civilisation, revolutions and utopias, and numerous other concepts and problems in social life'". A number of disciplines such as economics, history, sociology and jurisprudence have emerged to explain social life using scientific approaches based on the philosophical paradigm of post-positivism. The social sciences "are concerned with meanings, values, beliefs, intentions and ideas realised by human social behaviour and in socially created institutions, events and symbolic objects such as texts and images" (Harrington, 2005). The archaeological evidences and present built environment comprising landscape, structures and products have been seen as providing the 'data' or 'evidences' to understand social life using social theories. This also is a recent development.

The focus of architecture and built environment disciplines has been on the design and production of buildings, which has lacked concern on social determinants except as "background considerations related to site planning, public health and communal prosperity" until the twentieth-century departure from such a trend to some extent (Scaff, 1995). Since then, even though theories of architecture have acknowledged the "reciprocal influence between social forces and the organisation of space", ironically, "the theories of social and political life have little to say about architecture and community design" (Scaff, 1995). The absence of reflection on humanity's built environment is a serious shortcoming of social theories. This also reflects the differences in guiding conceptions for social theory and architecture. While social theory searches for "analytic epistemologies" and "grounded explanations", architecture theory is based on "normative compositional design knowledge" (Scaff, 1995). Over the past two decades, there have been attempts to engage social theory and architecture with the objective to find a vocabulary that is available to all (Scaff, 1995). The discourse, however, is still fragmented.

Criticising the narrow perspective of utilising archaeological materials as evidences for social interpretation, Fletcher (1995) argues that materials play a large-scale, slow behavioural role, which affects the viability of human communities. Besides human communities determining the structures and products of built environment, a reverse causality is equally probable, which poses limits on the activities that a settlement can perform given its built environment and limits on the size of settlements given the reach of social communication. To understand built environment of a city and its influences over space and time, we need a framework that integrates principles of social theory and related disciplines, architecture and planning. It may, however, be emphasised here that the objective of this section is not to derive a theory addressing these disciplinary shortcomings, which in itself is a work of diligence, but to determine a framework that could borrow principles from various disciplines to identify influences that can explain the built environment as it presents itself in present times.

Figure 1.2 presents an institutional pyramid to assist in understanding the evolution of built environment in Delhi and its surrounds. The pyramid is a generalised version, which allows discussion on the built environment at the city or regional level and is influenced by frameworks from Keogh and D'Arcy (1994) for property and Squires and Heurkens (2015) for property development. There are five layers of the pyramid, which are interlinked with each other. These layers influence each other in multiple ways. However, the distance between the layers reduces the degree of direct influence on each other. The bottom layer is the environments from which built environment evolves. These are the values, beliefs and norms of a society, which are reflected through its political, social, economic and legal institutions. These together form the governance structure, which influence all other layers of the pyramid. Moving up on the pyramid are the markets. Markets comprise of drivers and structures. What drives the demand and supply of

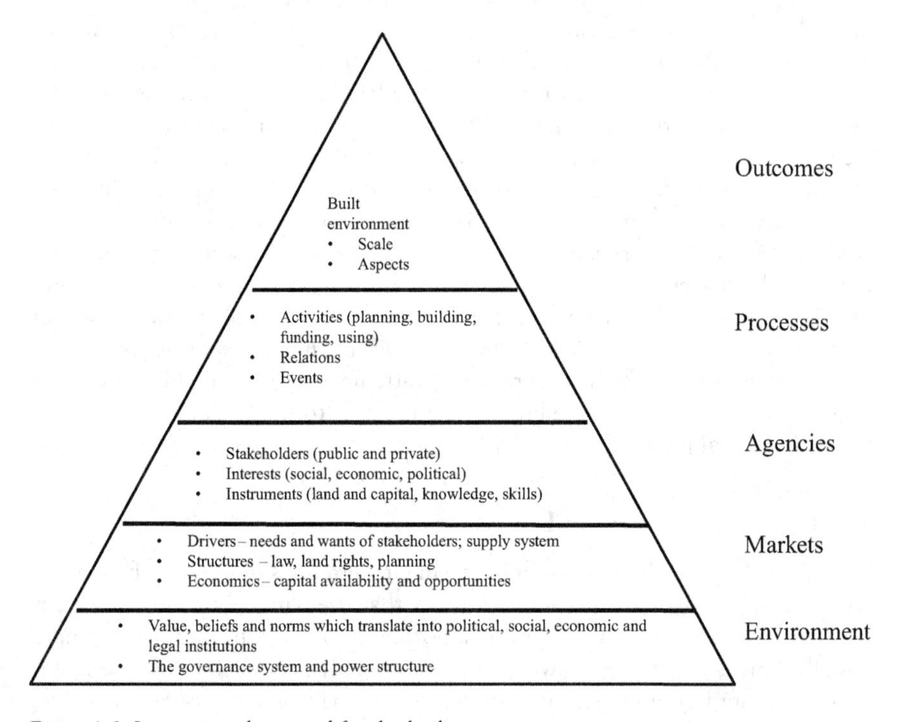

Figure 1.2 Institutional pyramid for the built environment

Source: Authors

built environment? Activities (e.g. education, business, residential, recreational, governance, storage of commodities, transportation, water storage systems) that take place in a city determine the demand and supply for the nature of space that is required. Some activities are strategic in nature (e.g. defence and security), which demand corresponding spaces (e.g. fortifications). The market structures such as the legal and planning systems and property rights etc. provide a mechanism for stakeholders to respond to market drivers in a holistic way. The capital availability and opportunities to deploy them among competing requirements provide what optimally would be built. The third layer is the agencies. These are the stakeholders who are involved in the conversion of natural environment into built environment. Operating with the social, political and economic interests, they utilise capital, land, knowledge and skills to develop built space. The fourth layer is the processes. These are the activities that take place in the development and use of the built environment such as planning, building, funding and using the built space. Processes that take place to shape the built environment are also determined by the relationships between those who perform these activities. Some or all of these activities are performed either by government and/or private sector. All these layers form the basis for what we see as the scale and aspects of built environment at the neighbourhood, city, regional, national and global levels.

The evolution of built environment as mentioned earlier is dynamic. Each of these layers has its own timeline of transformation. Environment, for example, takes relatively long time to change. This requires values, beliefs, norms of a society to change. This could happen with the change in political ideologies which change the norms for society and legal systems. If the new inherits elements of past, there is a continuity; otherwise fractures appear. In case of Delhi and its surrounds, history has witnessed many such transitions, which provides an interesting lens to see how these have impacted on the built environment. Transition time for markets is shorter than the environments, as these could change within the same environment over time. Agencies and processes have much shorter timeline in their transition as these respond to needs and priorities of society on a much regular basis. Outcomes reflect the impact of these different timelines. *Delhi's Changing Built Environment*, while examining the transformation of built environment, attempts to explain the forces that caused them. These forces are broadly the changes to the bottom four layers of the institutional pyramid.

1.4 Delhi and its surrounds

What have been the influences on Delhi and its surrounds that have left what we see as Delhi? Carl W. Ernst (p. 6) expresses that the "'influence' is nothing but a rather physical metaphor suggesting a flowing in of a substance into an empty vessel". This is rather a narrow perspective to apply to a city, which are metaphors of aspirations and ambitions of people who have lived or continue to live there. Understanding the transition and transformation of a cityscape is mired with complexities that transcend not only through time but also through the politics, society, culture and economics that contextualise the existence of a city. This becomes more complicated when the influences are also international as, then, a broad range of cultural manifestations emanating from another land – language, literature, concepts of governments, religious organisations, music and architecture, start to shape the space. Does the foreign start to collide with the native, or do these merge and create a unique identity? A city becomes a canvas of ideas, which gets imprinted on the spatial fabric as urban form and structures. It expands, it transforms, it assimilates and it becomes the crucible of human energy.

Delhi and its surrounds, as would be discussed in later sections, are in the words of Khosla and Rai (2005), 'the imagined conceptions of the rulers' and how the bureaucracies that were associated with these rulers implanted those imaginations on ground. Continuities of bureaucracies across regimes attempted to provide continuity in the built environment, sometimes seamless but at other times with rough edges. Evidences of these spatial interventions are the monuments and buildings spread all over the landscape of Delhi and its surrounds. The lattice of roads, streets, parks, canals and the River Yamuna that connects and intersects these monuments forms a mesmerising space.

Historians have often described 'seven cities of Delhi'. Others, by including New Delhi, have argued that an appropriate characterisation of Delhi is a city of 'eight cities'. This, of course, does not include the cities prior to 1100 AD for which built evidences are anecdotal though historical accounts and archaeological artefacts have been recovered.

Delhi has a long history, which despite occasional dislocation has shown a remarkable continuity and has the unique distinction of having been India's capital longer than any other city. Figure 1.3 depicts the eight cities whose footprints have dotted the landscape for more than a millennium.

Each of these cities grew around the palace-fortress of a particular dynasty, and every dynasty wished to have new headquarters that would be an epitome of prestige. Even the rulers of the same dynasty had these ambitions and realised them if they had the means to do so. With each successive reign, some distinctive architectural features were added or some change in urban morphology occurred.

The 'surrounds' are viewed as a dynamic boundary of the city to which it expands. It is ever expanding spatial expansion of existing city caused by the forces such as demographics, economics, planning, policy and the market. In the case of Delhi, 'surrounds' have changed over time as each new city formed.

1.5 The dynamics of political, economic and social context

Within the limitations of available information, this section recreates the social, economic and political profile of the ancient cities in North India, which also include Delhi. The historians acknowledge the dearth of appropriate archaeological explorations, which limit our understanding of the physical form of ancient Indian cities (Thapar, 2002, p. 119). However, it may be imagined how these cities would have reflected, in their built form, the social stratification, economic wealth and political supremacy and instability.

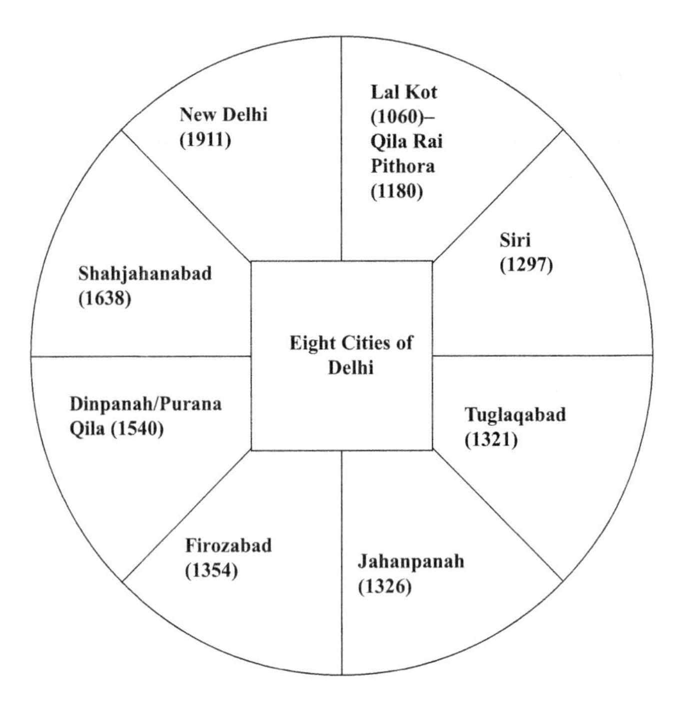

Figure 1.3 The time wheel of empires and the cities that they created

Source: Authors

1.5.1 Pre-1100 AD Delhi

The administrative boundaries of cities and states as we see now have undergone several changes over time. To explain more, historically cities like Delhi have formed part of bigger kingdom, of which the political structure and administrative boundary changed promptly due to chronic invasions and expansions, respectively. Therefore, it is important to study Delhi in association with the empire of which it has formed a part at different points in time. The study of the evolution of the Indian society tells that the formation of political institution is a complex phenomenon triggered by the demand for social organisation, economic facilitation and religion propagation. The following section discusses each of these drivers, in the context of North India, which together gave rise to political organisation.

First urbanisation in India – Indus Valley Civilisation

The northern region of India boasts of housing the Indus Valley Civilisation[1] as the 'first urbanisation' of India (Thapar, 1987) that dates back to 2500 BC.[2] There are continuous archaeological progressions found in the region which are indicative of the shift from the early form of village economy up till mature form of settlement in Indus Civilisation (Ratnagar, 2002). The general laws of history and archaeology suggest that "urbanism is not possible without a state level of political organisation" (Ratnagar, 2002, p. 8) and there is acute curiosity about polity of Indus civilisation. However, the reason why Indus Civilisation took a particular form, in reference to its mature economy and urbanisation, could not be explained properly due to the dearth of information, and little is known about the political structure of the civilisation, methods of mobilising surplus and ways in which the relationships were established and managed with other regions.

The Indus Civilisation started declining in the later second millennium BC (Ratnagar, 2002). Once again, the changes in political and economic systems are proposed by Ratnagar (2002) as possible reasons for the desertion of settlements in the Indus Valley Civilisation. The other reasons for decline, although weakly explained, include the flooding of the Indus River, lowered sea level, deforestation, tectonic movements, increased aridity and desertification, invasion by Aryans and similar others (ibid). Lack of information and unclear understanding of Indus civilisation has also created confusions about successor cultures of this mature civilisation (ibid). Geographical differences between the Indus and Ganges valleys, such as changing river courses, and climate change towards dryness in the Indus Valley are also brought forward as reasons for shifting of settlements into the Ganges (Thapar, 2002).

'Aryan' invasion?

The rise of 'Aryans' in the early first millennium BC was a simultaneous phenomenon to the decline of the Indus Valley Civilisation. The earlier hypothesis suggested that Aryans were foreign tribes who invaded North India and destroyed the established civilisation of Indus. However, this theory of invasion by Aryans is discarded by later historians who claim that there are no archaeological evidences which support the theory of large-scale conquest by a foreign culture (Thapar, 2002, referring to Possehl, Jarrige, Srivastava K.M, 1979, and Renfrew). Even the meaning of the word 'Aryan' is at times interpreted in a fashion that it supports the theory of foreignness. Thapar (2002) emphasises on the literal meaning of the term "Aryan", as mentioned in the Rig Veda, to be the people who are of respectable status and can use the Sanskrit language correctly, perform rituals and worship the right gods. Thapar (2002) insists that even though the Harappan culture was different from the Aryan society described in the Vedic literature, this does not necessarily infer that Aryans were foreign invaders. "The notion of Aryans being a physical people of a distinct biological race who moved *en masse* and imposed their language on others through conquest, has generally been discarded" (Thapar, 2002, pp. 95–96). Instead, the geographical progression from Indus in the west to Ganges in the east is often attributed to advancement achieved in iron technology, including the use of iron ploughshare for cultivation, which made possible clearance of flat and swampy plains of the River Ganges (Allchin & Allchin, 1968).

Beginning of settlement in Delhi with the 'second urbanisation' (or Ganges Civilisation)

The Aryan settlement in North India emerged and developed between 1500 BC and 500 BC on the other side of the watershed,[3] along the River Ganges, and is often referred as 'second urbanisation' or 'second civilisation' or 'Gangentic civilisation' in the Indian history (Thapar, 2002). As mentioned earlier, early historians believed that this new civilisation, established by Aryans, was independent of the Indus civilisation. Ratnagar (2002) strongly challenges the notion of this latter settlement developing in vacuum, disconnected from the former. Works by archaeologists A. Ghosh and B. B. Lal also reveal the possibilities of the succession of Indus civilisation, as post-urban phenomena, towards eastern side at Hakra, Sutlej-Yamuna divide, Punjab, Ganga-Yamuna Doab, Kathiawad and north Gujarat and even the Tapi valley (Ratnagar, 2002). Allchin and Allchin (1968) recognise the new evidences which indicate the extension of Harappan culture into the western Ganges basin (or Doab) stretching from present-day Delhi to Allahabad (in Uttar Pradesh, India). Similar discussions are presented by Thapar (1987), who writes that fresh evidences are found to indicate that the later civilisation bears many impressions of the Indus civilisation in spite of the difference in physical location. In a comparative way, Thapar (1987) writes about the technological advancement of this new urbanisation which was based on the use of iron, domestication of horses, the extension of plough in agriculture and a much more sophisticated market economy, thus indicating a progress ahead of the 'first urbanisation'. In conclusion, after the decline of the Harappan cities, many archaeological cultures succeeded them in various parts of North India (Thapar, 2002). Some of these cultures had overlapping elements of the Late Harappan civilisation, and many successor cultures had an independent genesis of their own (Thapar, 2002).

The pace of second urbanisation

The second civilisation of Ganges is usually divided into three geographical parts – western, central and eastern regions, which also correspond to distinct cultures that evolved over time, as settlements progressed from the west to the east (Allchin & Allchin, 1968). The western region covers Harappa in the Punjab, dry beds of the River Ghaggar in Rajasthan and the western part of Ganges-Yamuna river plains, between Delhi and Allahabad, called the Doab (Allchin & Allchin, 1968); the central region starts from the junction of the Rivers Ganges and Yamuna in Allahabad and covers eastern Uttar Pradesh and parts of Bihar state; and the eastern region is the delta of the Rivers Ganges and Brahmaputra, covering West Bengal (Allchin & Allchin, 1968).

It is important to mention the sites of historical importance in each of these regions. For example in the western region, including Doab, are located Kurukshetra and Hastinapur of Mahabharata,[4] Panipat, Indraprastha (or the Purana Qila mound at Delhi), Sonipat, Mathura and Bairat (Allchin & Allchin,

1968, p. 211). On the basis of literary sources,[5] it is believed that the city of 'Indraprastha', which was settled by burning the forest of 'Khandava-vana' in the first place, demarcates the beginning of settlement in Delhi. Lal (referred in Allchin & Allchin, 1968, p. 212) highlights the fact that many places mentioned in Mahabharata have been the settlements in this western region of Ganges.

The settlements in Doab gradually expanded into central plains of Ganges, which was otherwise covered by dense forests. This expansion was driven by increasing population of the Doab region and the availability of effective methods of forest clearance (Allchin & Allchin, 1968). Historically important sites in this region include Buxar, Chirand, Kausambi, Prahladpur, Rajghat (old Banaras or Varanasi) and Sonepur.

Private property rights

The early Vedic society was primarily hunters and food gatherers who identified wealth (or property) with cattle. Kosambi (1956) explains the importance of cattle to the food-gathering society and writes that "generally this (or property) meant cattle, which were first herded for meat, later for milk-products and skins (soon used in exchange); finally used as a source of power in agriculture and transport" (p. 22). The importance of cattle subdued gradually as the society progressed towards cultivation of food, and this increased the importance of land and its produce (Kosambi, 1956; Thapar, 2002). Kosambi (1956) writes that land was the territory (of a tribe) rather than a property (of an individual) at the time when human societies were dependent upon hunting and food gathering. The possession of land, as private property, became important when regular agriculture came into practice (ibid). "With regular agriculture, cattle manure fertilised land quickly exhausted by older tribal slash-and-burn cultivation; so permanent occupation of a field became the norm, tending towards private property in land" (Kosambi, 1956, p. 23).

The peasant society was also developing new tools that enhanced productivity of land, thus producing a surplus. It is interesting to note that the earlier society was either consuming or destroying the surplus produced, in contrast to the later society that focused on producing surplus for trade and exchange (Thapar, 2002). With intentions to maximise benefits from the surplus produce of land, the 'Aryan' tribesmen started developing their own land parcels, thus breaking away from tribal obligations of sharing the resources and outputs with other members of the society (Kosambi, 1972). The changing attitude of the agrarian society was therefore demanding formalisation of individual property rights on cattle, land and its produce which were otherwise owned in common by the tribe (ibid).

Specialisation of trade

Production of surplus consequentially encouraged specialisation of other skills circa sixth century B.C (Wagle, 1966). Childe (1950) explains that the production of agricultural surplus allowed exchange of food grains for specialised services from craftsmen, bricklayers (in Indus), transporter, merchants, officials and priests. This encouraged a few citizens to pursue full-time employment in non-agricultural trades. These specialists did not grow their own food and were rather dependent on the surplus produced by peasants from the city and dependent villages. The diversification of economic activities gave rise to new trades, thus changing the social composition and economic functions of the urban population. Wagle (1966) gives a detailed description of specialised occupations that emerged in North India at the time of Buddha (566 BC–486 BC[6]), such as trading; service occupations like washerman, barber, tailor and weaver; artisans; entertainers; professions like medicine, accountant and money-changer; king's services like warrior servants, archer, elephant rider, chief of the army; and other services like barber, cook who were employed for the king (the political structure of the society and the meaning of 'king' in ancient India's context will be discussed later). Wagle (1966) suggests that the political power and social prestige of the king was enhanced from the important role he used to play in the economy, being the single largest employer of multiple varieties of specialised persons.

Formation of guilds

The new groups of specialised craftsmen and artisans were usually kinship groups who later organised themselves to form guilds (*sreni*) (Kosambi, 1972). Trade and commerce were continuously evolving, and by the sixth century BC, traders emerged as the wealthiest and important class (Kosambi, 1972). The guild institutions often possessed considerable wealth at the disposal of the guild head or the guild council who at times gave loans to individual members of the guild or even externals (Kosambi, 1972). Traders were inviting huge social importance, to the extent that the financer or banker, who was at times also the head of the trade guild, was addressed as *sreshthi* which means 'superior' or 'pre-eminent' (Kosambi, 1972; Thapar, 2002).

'Specialisation' and stratification

Tribal settlements emerged as units of production of tradable goods. Thapar (2002) finds textual reference to villages of craftsman, such as carpenters and potters. Kosambi (1972) mentions Banaras (or Varanasi), where the entire village specialised in a particular trade like basket-making, pottery, smithing, weaving and other similar crafts.

Specialisation of trade also encouraged artisans associated with a particular craft to live in the vicinity so as to facilitate the obtaining of raw material and selling the finished goods to the merchants (Thapar, 2002). "It may also have developed out of caste considerations where occupational groups congregated; although it could equally well be argued that because of this congregation, some occupational groups came to be identified as separate castes" (Thapar, 2002, p. 124).

Diversification of economic activities in the society was giving rise to social stratification, although the division of labour was not rigidly crystallised, as in the later caste system, and people were able to change their occupation (Wagle, 1966). The two distinct kin groups which emerged during this time were the *brahmans* and *kastriya*. "The *brahmans* were ritually superior to the members of the ruling extended kin-groups or *kastriyas* but were politically subservient to them" (Wagle, 1966, p. 156).

Trade over long distances

The cities of seventh and sixth century BC were still small, and there was limited possibility of exchange with the neighbouring settlements (Kosambi, 1972). Surplus production and specialisation demanded long-distance trade with multiple markets on the way. Initially, the circuits of exchange developed over existing networks of pastoral groups (Thapar, 2002). However, this evolved over time, and new routes were explored by merchants who would travel in caravans of 500 oxen or more at a time, carrying tradable goods essential for day-to-day lives of that time like clothes, metals, bamboo (used as a building material), sandalwood (as a cleanser and coolant in the hot weather) and salt (Kosambi, 1972). Often, 'Kshatriya' tribesmen offered their security services on hire and escorted the merchant caravans (Kosambi, 1972). The merchant-caravaneers stretched over long distance from Taxila in the west to Magadha in the east (Kosambi, 1972, p. 124). Two important trade routes which developed were the *uttarapatha* (meaning the north route) and the *dakshinapatha* (meaning the south route). The soil was soft in the *uttarapatha*, and carts could be used easily for carrying goods. The *dakshinapatha* was a difficult hilly route with stony ground and broken passes, and often, trains of animals and porters were employed for carrying trade items.

Emergence of markets and formats of exchange

Various formats of exchange were developed by the market, including coins, letters of credit and promissory notes. This especially facilitated emergence of markets and encouraged trade over long distance for which traders could not travel themselves (Thapar, 2002). The units of measure and coins were already in use since 700 BC (Kosambi, 1972), and other formats were introduced as the markets matured. Thapar (2002) writes that the emergence of *sreshthi*, as a financer, was not merely the result of concentration of wealth and rather the changing nature of exchange, and trade also had a major role to play in their growth, as major stakeholders of financial institutions.

Blank silver coins were originally issued by traders themselves, and the guild institutions had an important role to play as guarantors of weight and purity of the coins (Kosambi, 1972). The coins were checked periodically during the period of circulation, and a punch mark was installed on one side to indicate the check made by the guild, to all who knew the "code of guild marks" (Kosambi, 1972, pp. 124–125). Similar style of checking and coding extended beyond the north route (or 'uttarapatha') of trade with Afghanistan, Iran and presumably even on

Achamenid darics such as those found in Gandhara (ibid). There is also connection with the Indus, and some of the punch marks are derived from Indus characters (ibid). Around the sixth century BC, political institutions became stronger, and the kings started putting their symbols on the other side of the coin that had once remained blank (ibid). As will be discussed later, Kosala and Magadha emerged as two major kingdoms, of which Kosala exclusively used four marks on its coins, while Magadha and the other remaining kingdoms used five marks (ibid). Thus, coinage was also symbolic of the kingdom and the empire where it circulated, and this trend is inherited by the modern currency systems as well, although the value of the currency is not directly linked with its material and is rather decided by a complex mechanism.

Efficient exchange of goods and services required social organisation (Kosambi, 1956). Traders had compressing needs for expansion of trade into other territories, and this required safe trade routes, establishment of harmonious relationships with people in other territories and standardisation of trade practices (Kosambi, 1972).

Political organisation and the society

The popular theory among historians is that the advancement in iron technology, and the introduction of the use of iron ploughshare caused significant improvement in agricultural productivity, thus producing surplus and helping market economy; these developments in turn transformed tribal societies into state system and provided the base for urbanisation (Thapar, 2002). Lal (1984) and Thapar (2002) consider these factors important, but insufficient to explain the conversion to state system. It is rather argued that the evolution of administrative and political system will be a rather complex process which cannot be explained fully by a single factor. The discussions in earlier sections analysed the social and economic changes, which together influenced the political system, as will be discussed next.

Basham (1969) paints a picture of the early stages of human settlements when the mankind lived truly in the natural state where there was no need of clothing, private property, government or laws. As and when man became more "earthbound" and institutions of private property and family took shape, they entered into fights, thefts and other crimes, to prioritise their own individual interests (Basham, 1969, p. 83). The tribal society then came together to "appoint one man from among them to maintain order in return for a share of produce of their fields and herds" and the "the great chosen one" came to be called the *Mahasammata* in the literal sense (Basham, 1969, p. 83). The title of 'Raja' was chosen for the elected person as it means the person who pleases everyone (Basham, 1969). Basham's reasoning takes inspiration from the popular theory of the English philosophers Thomas Hobbes (1588–1679) and his contemporaries. In his book *Levianthan*, first published in 1651, Hobbes (2010) discusses the arrangements which people made in order to protect themselves and their belongings: (i) *Pactum Unionis* – By this pact, people formed unions or societies and agreed to respect each other and live in harmony so as to give and receive protection of life and property to and from the members of the society; and (ii) *Pactum*

Subjectionis – By this pact people agreed to obey an authority and surrendered the whole or part of their freedom and rights so as to avoid conflict of interests (Hobbes, 2010).

It may be argued that the existing structure of the tribal society provided the platform for later development of a mature state system. As mentioned earlier, the tribes were often headed or controlled by a chief, called the 'raja' (Basham 1969; Thapar, 2002). The territory of a tribe was called the *janpad*, meaning 'foothold of a tribe' (Wagle, 1966). Tribal territorial units (or *janpads* or *gramas*, meaning village) were often named after the tribe (or *janas*), settled thereon, such as, Cedi, Gandhar, Kasi, Kuru, Kosala, Magadha, and Matsya (Thapar, 1987). This meaning of *janpad* was soon replaced by the new meaning of a kingdom or *rashtra* where many small tribal territories were agglomerated to form bigger *janpads* or *rashtra*[7] (Kosambi, 1972). The villages of a kingdom would come together only for common ceremonies and to fight against common enemies (Kosambi, 1956). The king was usually selected by election or by rotation from the whole set of tribal oligarchies (Thapar 2002; Kosambi, 1972). Among various other tribes in a *janpad*, the kastriya tribe were given the political control and land ownership. This arrangement gave rise to an oligarchic political organisation of ksatriya in most *janpads* (Thapar 1987). Lineage became an important guiding factor for allocation of political power, and kastriya tribe later became the royal lineages (Thapar, 1987).

Thapar (2002) explains that "the ability of the kastriya was based on his prowess in raids as a result of which wealth was acquired, as well as his ability to conquer territory to enable the establishment of new settlements" (p. 111). Attractive centres of trade and commerce, culture and luxurious living emerged along the River Ganges (Paul, 1996) which were worth plundering and capture. This created frequent disputes among different territorial units (Paul, 1996). Tribal identity was now giving way to territorial identity (Thapar, 1987). Due to constant increase in frequency of fights the power of the king increased and got restricted to one kastriya family (Kosambi, 1972). It was the time of Buddha (566 BC–486 BC[8]) when individual kastriya families emerged as powerful administrator owning large estates (Thapar, 1987; Kosambi, 1972).

There existed sixteen principal territorial powers between seventh and sixth centuries 600 BC, namely Anga, Magadha, Kasi, Kosala, Vajji, Malla, Ceti, Vamsa, Kuru, Panchal, Maccha, Sursena, Assaka, Avanti, Gandhara and Kamboja (Wagle 1966; Kosambi, 1972; Paul 1996). Kosambi (1972) mentions the "final struggle for power" (p. 121) that was fought towards the end of sixth century BC, post which four *janpads* retained the importance, which are Licchavi or Vajji; Mallas; Kosala; and Magadha. The former two were powerful tribal oligarchies, while the latter were absolute monarchies.

Forceful expansions and invasions certainly required a follow-up procedure of protecting the people so settled and wealth so acquired (Thapar, 2002). Growth of population and increasing prosperity and productivity were also demanding better administration that can provide protection and facilities to the residents. To fulfil these requirements, each territory maintained a permanent army and collected revenues, which were used for maintaining the army and for expansion of territories, construction of highways and public works (Paul, 1996). The gahapatis, traders and farmers were the important stakeholders of the society who contributed heavily to the state revenue and who were redefining the needs of the new society (Kosambi, 1972). As discussed earlier, two important changes were demanded by these stakeholders – (i) formalisation of private ownership of land and its produce (Thapar, 2002) and (ii) safe trade routes and establishment of harmonious relationships across territories that can facilitate trade across territories (Kosambi, 1972). These requirements were conflicting with the tribal norms in many ways; for example, the norm of redistribution and communal ownership did not allow individual property rights; and the tribal chiefs would normally not submit to regularise taxes, which were otherwise necessary for generating revenue, using which a permanent army could be maintained (Kosambi, 1972). Therefore there emerged a need for a new political system that could establish central control of a universal monarch who can break through the rudimentary tribal administration and develop new policies and institutions (Kosambi, 1972).

This was also the time when Buddhism was developing as a new religion, and Kosambi (1972) relates the development of "universal monarchy" to paralleled development of the "universal religion". In this context, Kosambi (1972) writes,

> The new moral philosophies of the sixth century B.C. that formulated and preached a doctrine beyond the tribe had their political counterpart. There was a parallel move towards a universal government for all society. The basis was identical both in religious and secular movements: the new needs of gahapati, trader, and farmer. Whereas the founders of the great monastic orders had considered tribal patterns of organisation quite suitable and perfectly natural for their own samghas, especially the Jain and the Buddhist, the theoreticians of state policy could think of only one way to break tribal exclusiveness – a dictatorial absolute monarchy.
>
> (p. 120)

Founding principles of administration and economy were then laid in 'Arthashastra', by Chanakya, and state system was crystallised during the Mauryan age (273–32 BC) (Sharma, 2005). Henceforth, a strong foundation was laid for further development of politics and economy. Later, during the Gupta age, the polity, economy and judiciary matured, thus giving a strong foothold to these rulers who covered much of the Indian subcontinent between 320 AD and 550 AD. Stable governance stimulated the development of trade and commerce, and the growing prosperity of the country invited attention from foreign invaders, who eventually made their way and founded a new religion, a new system and also new cities in India.

The period from the post-Gupta centuries to the establishment of Delhi Sultanate in the twelfth century, referred as the early medieval period, was transitionary where the country was divided among regional powers and lacked political unity. Kulke and Rothermund (1999) argue that "this absence of political unity contributed in many ways to the development of regional cultures which were interrelated and clearly demonstrated the great theme of Indian history – unity in diversity". The political system that emerged during this period was the feudal system, emerging from the changes in the nature of politics and material culture. Sharma (2006) describing the 'Indian Feudalism Model' describes the feudal characteristics of early medieval periods as "the increase of religious intermediaries in land, the payment of lesser vassals and officials by land grants, the feudalism of titles of kings and officials, the shifting of capitals, the imposition of clan chiefs on old villages". While empires of Mauryans and Guptas were highly centralised, the medieval India had multiple centres of power (ibid). The practice of land grants with transfer of administrative, fiscal and judicial powers was a remedial step to counter the problem of difficulties being faced in surplus extraction. This decentralised power (ibid).

Chattopadhyay (2012) views the above formulation of feudalism too simplistic as it merely views resource transfer centre to the feud as the means for creation of feudalism. While he does not question the feudal system, he proposes that it was the result of complex interplay of political, social and economic system of the time and space. Chattopadhyay (2012) argues that the emergence of feudatory system can be explained on the basis of lineage power of ruling elites. The transformation of lineage into regional power was a result of support from other lineages over time, won over by land grants and conferment of feudal ranks. Kulke (1999) identifies three stages: first, tribal chieftain would turn into a Hindu prince, then this prince would become the king surrounded by feuds. This would establish 'early kingdom', and in the third stage, great rulers of emperial kingdoms emerged who would control large realm and integrate feudal lords into internal system of their realm.

In North India, Gurjara-Pratiharas emerged as independent dynasties in different parts with their origin explained largely with the theory of Indian feudalism mentioned earlier. The Pratihars who established regional kingdoms called themselves Rajputs. Land grants that were offered usually extended in areas that were outside the already-cultivated areas, which while allowed allegiance of feuds also expanded the territories of the kingdom. Battles among Rajput kings were common. One of the great Rajput warriors, Prithviraj Chauhan, was at the helm of affairs in Delhi when Mughal invaders arrived in Delhi in the twelfth century.

1.5.2 The period of Muslim rulers

This period of about 750 years spanning over numerous dynasties presents an important phase in the development of Delhi. During the first half of this period, the rulers, who were foreigners and had very different religions and cultures, were trying to establish themselves in this new land. While the rulers of the second half of this period also came from foreign lands, they had the advantage that their predecessors had made significant attempts to bring religious, social and cultural congruence at least in the areas where they were seated.

Politics

The eleventh-century attacks by Gazanavids on India were largely with the intention of amassing wealth to provide means to invaders to play an active role on the Persian plateau. However, with the defeat of Gazanavids in Persia the epicentre of their power shifted to the Punjab (north of Delhi), and Lahore became the capital of the remaining empire of Gazanavids, which once commanded vast areas of Persia, Afghanistan, much of Transoxiana and northwest part of Indian subcontinent with capital at Ghazna (in Afghanistan). With the conquest of Ghazna by the Ghurids' Turkish *gulams* (slaves) in 1150, it marked the end of Ghaznavid rule west of the Indus. In 1186, the Ghurid ruler Ghiyath al-Din Muhammad occupied Lahore, where he established his rule together with his younger brother Mu'izz al-Din Muhammad, to whom he delegated the eastern and southern possessions of the dynasty. Thereafter, Mu'izz al-Din Muhammad was responsible for the extensive conquests in India. Delhi was captured in 1192, Ajmer in 1193 and Kannauj in 1198; Ghurid suzerainty thus extended in a great arc from Mount Abu in Rajasthan through Gwalior to Bundelkhand. Following the assassination of Mu'izz al-Din Muhammad in 1206, his territories were partitioned among his principal nobles. A slave, Quṭb al-Dīn Aibak, became viceroy of Lahore, Ajmer and Delhi, the last city being held by his lieutenant Iltutmish. Eventually, Aibak emerged more or less supreme, but he never assumed the title *sultan*. After his death, his successor was his slave and son-in-law, the far-sighted Iltutmish, who is counted the first and among the greatest of the sultans of Delhi.

The Turkish slaves of the Ghurids who laid the foundations of Muslim rule in India were no barbarian conquerors. Rather they were educated who became effective transmitters of Persian civilisation on the subcontinent. The earliest surviving buildings erected by the sultans of Delhi reflect an interesting mixture of Indian and Persian styles. The Ghaznavid and Ghurid invaders constituted a well-defined ruling elite, reinforced by adventurers of all kinds from the Muslim lands farther west. The prospects of large booty from wars in Indus region attracted many men from Khorasan, Transoxania and Sistin to join the army (Nezam-al-Molk, p. 146 as cited in iranicaonline). Apart from soldiers, little is recorded about early migrants from Persia and the borderlands into what later became the Delhi Sultanate. It can be assumed, too, that among immigrants to northern India there were *ulamas* (Muslim religious scholars), armorers, metalworkers, tentmakers and furnishers, manufacturers of cavalry gear and other craftsmen, though none is mentioned in the sources. Merchants must have followed the armies to convert the plunder into cash; the vast majority of Indian captives must thus have become objects of commerce. Traders and craftsmen alike most probably came from urban centres in the eastern Persian world and, with bureaucrats and *ulamas*, provided the nucleus of the free, non-military Persian-speaking population of such centres as Multan, Uch, Bhakkar, Lahore, Dipalpur and Bhatinda in the Punjab, as well as Delhi.

Politically, this period was turbulent with frequent changes in rulers. Usurpation and murder more often determined the succession at Delhi. Iltutmish was succeeded by five descendants. In 1266, the slave Giat-al-din Balban captured the throne, ruling for two decades in unpleasant extravagance and with utmost centralisation of power. After his death his grandson and great-grandson were soon ousted, and the throne was then seized by the Turkish Ḵhiljis (1290–1320), who founded the second dynasty of Muslim rule in India. Though their rule of three decades was of high level of centralisation of power, they are credited for their contribution to architecture and built space in Delhi. In a major shift from earlier Muslim rulers who had ruled over Delhi from the forts and city constructed by their Hindu predecessors, Khilji laid the foundation of Siri – the second city of Delhi, but first originally built by a Muslim ruler in India. As will be discussed in Chapter 4, a number of structures, architectural features and styles in Delhi are credited to Khilji rulers.

After the assassination of the last Khilji ruler, Gazi Malik, governor of Dipalpur (Punjab), ascended to the throne as Ghiyath al-Din Tughlaq and founded the Tughlaq dynasty (1320–1414). During this period, the sultanate of Delhi reached its greatest extent but also experienced the beginning of fragmentation into smaller states. The rulers of Tughlaqs, especially Muhammad bin Tughlaq (1325–51) and Firuz Shah (1351–88), brought political stability and the Delhi Sultanate reached the zenith of its splendour. Tughlaq rulers built Tughlaqabad (third city of Delhi, later abandoned due to shortage of water), Jahanpanah (the fourth city between Qila Rai Pithora and Siri) and Firozabad (fifth city of Delhi).

The decade that followed Firuz Shah's death saw Tughlaq Empire fragmenting under the weak rulers who had followed. Even before Timur's devastating raid on Punjab and Delhi in 1398–99, the Tughluq state had contracted to little more than the countryside immediately surrounding Delhi. Timur's raid further weakened the state, and he installed his loyal Sayyed Kezr Khan on Delhi's throne in 1414. The last ruler of the Sayyed dynasty abandoned Delhi in 1451 for Badaun, and Delhi was taken over by Bahlul Lodhi, forming Lodhi rule in India. Lodhis, who ruled for three-quarters of a century (1451–1526), were themselves overthrown by Babar, thus sowing the seed for Mughal Empire in India. The most significant legacy of the Sayyeds and Lodhis was architectural, as discussed in Chapter 4.

Babar founded the Mughal Empire in India in 1526 after he defeated Lodhis in the Battle of Panipat, near Delhi. After a brief intermediate rule of Sur dynasty from 1540–55, Mughal ruler Humayun recaptured India. He established himself at Purana Qila, the fort that Sher Shah Sur had constructed in Delhi. Humayun founded the city of Dinpanah, of which Purana Qila was inner fortress. Akbar who inherited the throne after Humayun's death expanded and consolidated the empire. The political, administrative and military structures that he created to govern the empire were the chief factors behind its continued survival for another century and a half until the death of Aurangzeb in 1707 (Editors of Encyclopaedia

Britannica, 2017). Mughals ruled India for more than three centuries though the empire weakened after Aurangzeb's death due to weak rulers and infighting for succession, but survived for another 150 years. Though Delhi was not always the capital of Mughal Empire as the imperial power shifted from Agra to Delhi to Agra to Fatehpur Sikri to Lahore to Agra and back to Delhi, it remained an important city during Mughal rule (Sinopoli, 1994). For about the last two centuries of Mughal Empire, Shahjahanabad remained the capital until 1857, and no new city in Delhi was formed.

Blake (1991) characterises the Mughal Empire as a patrimonial bureaucratic system, with its political and economic structure revolving around the royal household and the succession on hereditary lines. In his research on Shahjahanabad, Blake has viewed that the capital city constructed by Emperor Shah Jahan in 1631 is the patrimonial household on a larger scale. This perspective highlights the significance of the Mughal emperor and his imperial household in the polity of state (Sinopoli, 1994). As discussed in Chapter 3, Shahjahanabad was a well-planned city with properly planned markets, cantonment, places of worship and gardens.

The period from 1748 onwards was turbulent for Mughal Empire. Frequent raids by Afghans from the northwest between 1748 and 1767 and the peace treaty with Marathas in 1751–52, which effectively constricted the Mughal emperor's authority to Delhi, and the later surrender to the British forces in 1803 reduced the status of emperor merely to a nominal sovereign on pension with no real authority. These arrangements continued with the later rulers, and their dependency on British generosity kept on increasing.

Though the British East India Company's capital was Calcutta (Kolkata), Delhi was important, as it was the seat of emperor. The local officials during the later Mughal and Maratha period were corrupt and less hesitant in pursuing their private interests. With control, British made changes to the land revenue administration system by moving towards permanent settlement system that they had implemented in Bengal (see Chapter 2), payment of land revenue in kind instead of cash and attempts to eliminate middlemen from the collection of land revenue through various arrangements discussed in Chapter 2.

Politically, Delhi was merely a military outpost of British East India Company till 1857. Prior to building new army cantonment near the north ridge, British had considered Daryaganj as the site for the cantonment largely due to its proximity to Red Fort. However, subjugation of Mughal emperor to British authority made them reconsider their decision regarding the location of cantonment away from Daryaganj. The area around the north ridge was developed as British outpost in Delhi, as discussed in Chapter 3, in a rather haphazard manner.

Economic development of Delhi

The economy of Delhi before Sultanate period was largely agrarian. Peasants lived in villages of 200–300 men and carried on agriculture. Size of farms differed. Headmen had large landholdings, while village menials had petty plots. There was probably a large landless population, about whom information is sparse. Delhi was largely occupied to the north of the River Yamuna, as there are descriptions of dense forests and tigers in the area between Delhi and Badaun or in the middle Doab region. This land was largely uncultivated, which of course changed by the end of the sixteenth century.

It appears that dense vegetation occupied the Doab region (between the Rivers Ganges and Yamuna) and was probably difficult to clear for agriculture. Consequently, much of the agricultural activities concentrated on the northwest region of the River Yamuna. The region northwest to Yamuna is also connected to Rajasthan, from where earlier Rajput kings came to Delhi. Wells and canals were probably the sources of artificial irrigation.

Alauddin Khilji, the first Khilji ruler, was expansionist and needed resources to pay for his growing army and to meet the cost of wars. He changed the land revenue policies to strengthen his treasury to help pay his growing permanent army and fund his wars by raising agriculture taxes from 20% to 50% (Raychaudhuri, 1982), rationed food and prices, eliminated payments and commissions on taxes collected by local rajas (kings) and banned social contact and ties among his officials to prevent any opposition forming against him, and he cut the salaries of officials, poets and scholars (Hermann & Rothermund, 2004).

The taxation methods pursued by Khilji increased land revenue substantially but negatively impacted agricultural output. To combat rising inflation, Alauddin introduced price controls on all agriculture produce, goods, livestock and slaves (Jackson, 2003). He also designated markets called *shahana-i-mandi*, for the sale of goods and services (ibid). A permit system was used to allow merchants to buy and sell goods at official prices. These harsh measures had a detrimental impact on farmers which led to more incidences of food shortages in rural areas of Delhi Sultanate to which urban Delhi remained largely immune (Lal, 1967). With the death of Alauddin Khilji, the price rationing system collapsed, and Delhi witnessed huge inflation (Raychaudhuri, 1982).

In the words of Raychaudhuri (1982), "Alauddin Khilji's taxation system was probably the one institution from his reign that lasted the longest, surviving indeed into the nineteenth or even the twentieth century. From now on, the land tax (kharaj or mal) became the principal form in which the peasant's surplus was expropriated by the ruling class".

Another major reform to the land taxation system came during Sher Shah Sur's regime (1538–45). He got all land measured, and land tax was set as half of the produce. Tax liabilities were written, which eliminated the arbitrariness of earlier systems. The taxes were collected directly from peasants. While the system continued during the Mughal Empire that followed Sher Shah Suri, Akbar changed the system of determining land tax rate to cash basis, a system which remained intact until 1857.

Agriculture was important for the revenue of the Sultanate, and rulers did take initiatives to augment irrigation systems to bring more areas under cultivation and increase yield. The first ruler credited with digging canals for agriculture was Ghiyath al-Din Tughlaq (1320–25). Canals from earlier period were later dug by Firuz Shah Tughluq (1351–88) to promote agriculture. He undertook major canal building task connecting the River Yamuna to Hisar (about 160 km), in the northwest region of Delhi. Firuz Shah Tughluq's predecessor Muhammad Tughluq (1325–51) had advanced loans to peasants for digging wells. In his time, dams were built, some funded privately by people and others by the government. During the reign of Tughluq, agriculture advanced. Hisar was so well irrigated that instead of one crop, farmers were able to harvest two crops in a year. Two agricultural patterns are important from that period, one near the irrigation channels, which allowed intensive agriculture. Crops such as sugar cane were grown on well-irrigated land. Other land that depended on rain as source of irrigation could grow staple cereals (e.g. gram and barley). Kumar and Raychaudhuri (1982) have compiled the crop output and prices in different periods since 1290 AD. It becomes important to highlight that crops that depended on irrigation such as sugar cane fetched very high prices in markets compared to those that did not. Land was abundant. The land pattern that emerged was intensive cropping around sources of artificial irrigation and rain-grown crops on vast areas of land. There was also state encouragement for growing certain crops. Raychaudhuri (1982) presents evidence that Muhammad Tughluq urged peasants to grow grapes and dates. His successor, Firuz Tughluq, laid out 1,000 orchards of grapes in the vicinity of Delhi.

Another important aspect of economic life was raising cattle. As discussed by Raychaudhuri (1982), the large area of wasteland, including fallow land and forests, implied that there was shortage of pasturage for cattle. One account of villages during Firuz Shah Tughluq's period suggests that villages had 40–50 cattle-pens. Cattle were abundant, and, probably, households raised their own cows and bullocks. These animals were raised for production of milk and milk products. The manufacture of butter was important and profitable trade. Bullocks were also used for transporting grains.

The rural system remained more or less undisturbed during Mughal time. Rulers maintained the old organisations of villages, and often land revenue was collected at the village level rather than individual level. As will be discussed in Chapter 2, Mughals used three systems to collect land revenue – through intermediaries (zamindari system), directly from peasants (*Raiyatwari* system) and from village collectives (*Mahalwari* system). The system was organically developed rather than imposed by the rulers. Rulers also advanced loans for seeds, implements to dig wells, bullocks. Mughal rulers, in particular Akbar and Jahangir, encouraged trade. The efficient city administration facilitated trade, as has been articulated in the accounts of many foreign travellers of that time. Bernier in his travel memoirs during Aurangzeb's time writes that much of the commerce was in hands of traditional Hindu merchant classes, who, supported by caste guilds, had developed an efficient system of commerce. While Muslims were preferred for administrative posts, Hindus dominated trade and commerce. There are accounts of Aurangzeb's "wars of succession" and in Bengal which were financed by loans from Hindu merchants.

Social system and settlements

There is tentativeness to the records on social hierarchy that existed during pre-Sultanate and Sultanate period. Historical accounts (see, for example, Raychaudhuri, 1982) indicate that during the pre-Sultanate period there were three levels of administrative hierarchy (which also accorded social status) above peasants and village headmen (called *khots*) – Ranaka, Raja and Rauta. Ranakas were immediate vessels of rulers who were granted land by the rulers. They could collect taxes and could also assign land rights to others below them without permission from rulers. They were, however, obliged to provide their services during war and pay taxes to the grantor and protect villages from disturbances and sudras. Rajas were lower than Ranakas and received land grants from Ranakas in lieu of military services that they provided and taxes that they collected from land that was assigned to them. It appears that they could not bestow further land grants. Rautas were men of lower status than Ranas and Rajas and probably commanded a smaller army. This system continued with very little modification till the Sultanate period. While the plans, buildings and remains of the Sultanate period or later exist in some form in archaeological archives (as will be discussed in Chapter 2), the plan of a settlement in Delhi prior to that is a matter of guesswork based on the social system that existed. A possible settlement in Delhi during pre-Sultanate times may have looked like Figure 1.4.

Delhi was little affected by Islamic culture prior to defeat of the last Hindu Rajput ruler of the city, Prithviraj Chauhan, in 1192 by the Ghurid general Quṭb al-Dīn Aibak. By 1193 Aibak had conquered Delhi itself and had established Islam as the new state religion. Medieval Persian institutions, already established in Ghurid Afghanistan, were also implanted in Delhi (Centre for Iranian Studies, 2017).

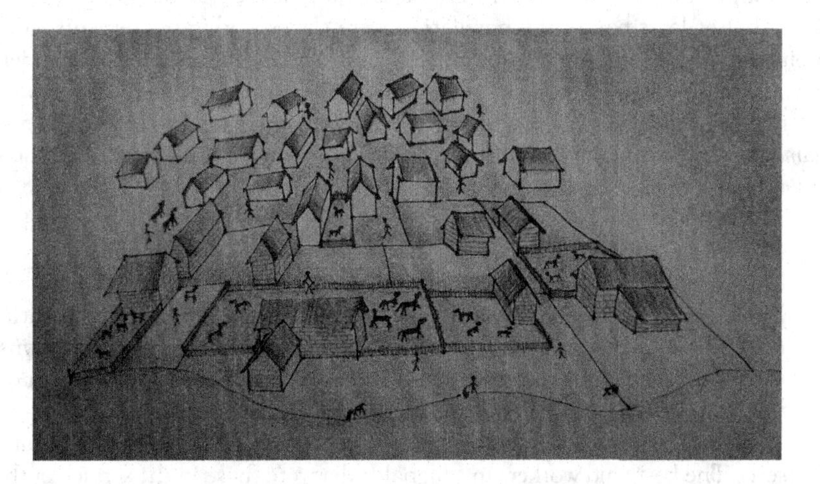

Figure 1.4 An artistic impression of a settlement in Delhi during the pre-Islamic period

Source: Authors

Although there are no formal records to confirm the extent of immigration into India before the 1220s, events during the thirteenth century provide circumstantial evidence. The first wave of immigrants came from the northwest as a result of campaigns by Genghis Khan in Transoxania and Khorasan in 1219–22 (Centre for Iranian Studies, 2017). Many fugitives sought refuge in Delhi during the rule of Iltutmish, which led to a greater penetration of Persian customs and values in lands that had previously seen insurgents looking to plunder India's wealth in an attempt to increase their influence in their homelands. The invasion of Persia by Mongols continued through 1250s and later, and it must be assumed that the exodus also continued, though presumably limited to persons of means or possessing marketable skills (Centre for Iranian Studies, 2017). Among the refugees who came to Delhi was the party with whom Ibn Batuta travelled in 1333 (ibid) who was part of entourage of Qazi of Termed̲. Sultan Muhammad bin Tughlaq was especially renowned for his hospitality to foreigners. The Centre for Iranian Studies (2017) cites Ibn Batuta's mention about the sultan's practice "of honouring strangers and showing affection to them and singling them out for governorships or high dignities of state" and that the sultan was "well known in his generosity to foreigners, for he prefers them to the people of India, singles them out for favour, showers his benefits upon them . . . and confers upon them magnificent gifts". The rulers were keen to follow practices of Persian kinship, as there were no local Rajput rulers' practices to looks forward to. It appears from these travelogues that the impact of Persian culture and customs started to reflect in everyday life of Delhi.

With the establishment of Muslim rule, the class system emerged in Indian society. At the top of hierarchy were the upper-class nobles and religious people in the state, mostly immigrants. Native Indians took lower positions in the social and administrative hierarchy. Slavery became a practice. Raychaudhuri (1982) notes that at least half of the population were slaves working as servants, concubines and guards for the Muslim nobles, amirs, court officials and commanders during Khilji's reign. There was a visible discrimination against those slaves who belonged to the poor classes in India. Ifthikar (2017) concludes that nobles and *ulamas* became very powerful with Iqta system grant of piece of land for some assignment after the establishment of Muslim rule in India.

In the words of Ifthikar (2017),

> Mughal society was starkly unequal. Money was circulated only in Urban India. In rural India there was hardly any circulation of money. The entire rural labour force was made up of virtual serfs who were merely fed. In towns, there were labour markets from where the workers hired for daily wages or on the contract. But that didn't make any difference in the economy. The wages of their labour were so low that their condition was little better than rural serfs. The best paid workers in Mughal India were those in the service of the Emperor. The wages of these workers were for the first time fixed by Akbar according to their skills. Mostly workers of that age drew monthly salaries. Daily wages were only confined to the lowest categories like sweepers, bamboo cutters, grass-cutters and helper boys in the royal stable.

During Mughal times, the artisan skills of Indian craftsmen were developed to an exceptional degree. According to Bernier (as cited in (Ifthikar, 2017), "handsome pieces of workmanship made by persons destitute of tools . . . sometimes they imitate so perfectly articles of European manufacture that the difference between the original and copy can hardly be discerned". There was the abundance of skilled labour working on low wages, largely divided on the basis of caste system to undertake occupations.

1.5.3 British Delhi

Originally a trading company with little interest in politics, the British East India Company, in a change of strategy, started playing a major political role after it struggled with its French counterparts in dominance of trade in India. The Battles of Plassey (1757) and Buxar (1764), which saw British troops defeat Indian powers and take custody of Bengal, a rich province of the weakening Mughal Empire, led the company to rethink its engagement in India. In the following decades, the company gradually increased the extent of the territories under its control, ruling directly or through titular rulers under the threat of force. The company started to exercise military power and assume administrative functions over a large territory in India. Delhi came under the administration of the British East India Company in 1803 when Mughal emperor Shah Alam sought British help in defeating Marathas. In theory (see Buckler, 1922), the company started to administer Mughal territories on behalf of the emperor, and its authority was derived from emperor's orders (*farmaans*). In practice, the emperor was mere a titular ruler with no real authority. Spear (1951), cited in Ikram (1964), writes that "[w]ithin the walls of the Red Fort the king (Mughal emperor) retained his ruling powers. The inhabitants of the Fort bazar were his direct subjects, and the members of the imperial family who lived within enjoyed diplomatic immunity. The etiquette of the court was maintained, the sonorous titles and the language of the great Mughals continued, and the (British) Resident attended the durbar (court) in the Diwan-i-Khas regularly as a suitor. He dismounted like any other courtier . . . and was conducted on foot . . . to the imperial presence where he stood respectfully like the rest". Outside the walls of the Red Fort, the authority of Emperor was nominal. The tussle between British and Emperor extenuated during the wars of succession and successive Mughal rulers' reigns. The continuous dominance of the British East India Company on administration, ensuing disagreements over the amount of pension owed to the emperor by the company and the title and status of heir apparent of Mughal emperor Bahadur Shah led to the emperor supporting the uprising against British in 1857.

The defeat of rebels in the hands of British armies led to the exile of Mughal emperor, which marked the end of the Mughal era in India. Ikram (1964) states that 'the entire population was driven out of the city, and in the absence of owners, the houses were broken into, their floors dug up, and contents removed or destroyed'. He writes further that the "next to suffer were the city buildings. The principal mosques were occupied by the British troops. One proposal was to sell the Grand Mosque of Shah Jahan. Another was to convert it into a barracks for the main guard of European troops. Muslims were not allowed to use it until five years later. Some parts of the Fatehpuri Masjid, the second largest in the city, remained in non-Muslim hands till 1875. The beautiful Zinat-ul-Masjid, built by Aurangzeb's daughter, was only restored to the Muslims . . . at the beginning of the twentieth century. The royal palace and the fort suffered even more. The palace proper, the residence of the royal family, was razed and all the gardens and courts were completely destroyed".

After the suppression of uprising, through the Government of India Act 1858, the control of India was transferred from the company to the British crown, and India came under the rule of British Empire. Initial seat of government was Kolkata but in 1911, the capital of British India rule was decided to be shifted to Delhi, and, as will be discussed in Chapters 2 and 5, construction of a new capital for British rule in India commenced.

British rule in India led to changes in education, social and administrative systems, with more involvement of Indians in governance. At the same time, the demand for "self-rule" by nationalists was growing. There was discontent with the approach and rule by the British, who were viewed as foreigners in India working in their self-interest without much concern for the development of India and its people. The society that emerged after 1857 was the one that was divided along religious lines. R. C. Majumdar (as cited in (Ikram, 1964), the Indian historian, has said, between Hindus and Muslims,

> the social and religious differences were so acute and fundamental that they raised a Chinese wall between the two communities, and even seven hundred years of close residence (including two of common servitude) have failed to make the least crack in that solid and massive structure, far less demolish it.

Nehru, however, blames the division in society on British policies. In his book *The Discovery of India*, he writes,

> the British Government had also stood in the past, in theory at least, for Indian unity and democracy. It took pride in the fact that its rule had brought about the political unity if India, even though that unity was one of common subjection. . . . But curiously enough its policy has directly led to the denial of both unity and democracy.
>
> (Nehru, 1981)

Further, he writes, "In August, 1940, the Congress Executive was compelled to declare that the policy of the British Government in India is a direct encouragement of and incitement to civil discord and strife" (ibid). The independence of India also led to its partition into two nations: India and Pakistan. In 1947, India became a democratic republic and free from British dominance and subjugation. Huge migration of humanity on religious lines on both sides of borders led to huge influx of population in Delhi.

1.5.4 Changing context for the capital of the Republic of India

During British rule, cities such as Kolkata and Mumbai were the engines of manufacturing and trade, largely producing processed raw material for the British textile mills. The British Empire had just got out of war, and Delhi had completed construction of a new capital when India got independence. At the time of independence the conditions were precarious. Various economic linkages such as production-market-trade, rural-urban interdependence, sources of investment that the country had were broken (Tiwari, 2014). The trade port of Karachi and the jute-manufacturing hub of Bengal were lost to Pakistan. The country faced serious economic challenges on the eve of independence and the decade that followed (ibid).

India witnessed phenomenal growth in population immediately prior and after independence. The earlier growth in population was largely led by industrialisation and fading opportunities in rural areas, which drew people from the hinterland into the cities, and post-independence, there was huge migration due to the separation of nations. The surge in population got concentrated in big cities – Delhi, Mumbai and Kolkata. The image of Indian cities at the time of independence, or just after independence, was very fragmented, as the economic base was getting lost, but there was a huge influx of people (Tiwari, 2014).

The political elite which governed India immediately after independence was Western educated and well aware of development paradigms elsewhere. As the first prime minister of the country, Nehru wanted to make India self-reliant by producing and investing in capital goods so that it didn't have to import all those things from elsewhere. In the minds of political masters at that time, strong control on foreign direct investment was an appropriate strategy, more so because of poor experience with the British East India Company. In a change from pre-independence period, though there was a focus on agriculture because India had a big mass to feed, policies were driven in the direction of capital good investment. Heavy manufacturing industries, which would produce capital goods, were the agenda for the political leaders (Tiwari, 2014).

To finance capital manufacturing investment, personal and government savings were channelised. The rural surplus became the driver for investments into heavy industry. To implement this vision of an independent India, it was thought that the policies should be driven from the centre in the form of Five-Year Plans, and a planning commission was set up, which was playing a role by formulating these plans (Tiwari, 2014).

During the 1960s and 1970s, as the democracy was gaining ground, discontent in rural areas was growing. It led to the emergence of local leaders who were against that urban bias in investment policies. They wanted India to focus more on the rural rather than basing on inefficient capital-intensive activities, which don't generate much employment. A few monsoon failures and droughts further escalated such sentiments. In 1962, India and China went on a war that aggravated the situation in rural areas further. Public sentiment, which was against the urban bias, was capitalised by politicians who had rural bases. The vote bank policies that followed led to a shift in focus towards rural areas. The agenda shifted to eliminate poverty rather than growth. It started hindering economic growth, and the neglect of cities began. The unproductive public expenditure, such as agriculture subsidies and investment in canal irrigation, began, which depleted the exchequer (Tiwari, 2014).

During the 1980s, India started to re-evaluate its approach towards development. During the late 1970s and 1980s, there was a shift in development paradigms around the world. China started pursuing its economic liberalisation policy post-1978. The Thatcher government in the UK and the Reagan government in the US instituted mechanisms to reduce public sector dominance. At home, issues such as labour unrest in Mumbai textile mills, which completely collapsed the textile industry in Mumbai, drove the desire to modernise. India needed capital and technological infusion. The world had become politically and economically stable, and it was realised that foreign investment is not that bad, so a process of cautious liberalisation began.

During the 1990s, India witnessed a catastrophic economic situation, as it faced a balance of payment crisis. The last four decades had also taught that urban sectors (manufacturing and services) were contributing far more to the economy than agriculture. This led to liberalisation in a big way in the 1990s, and most sectors were opened up for private investment. Despite all this, cities that had been ignored faced challenges related to poor living conditions and housing shortages. Government attempts, as discussed in Chapter 2 for Delhi, largely failed to house migrants who came to cities as rural economy stagnated. Slums mushroomed all over the landscape. Slums in these cities are functioning entities where people live, work and socially intermingle. Unlike slums in the Western world, wherein these are largely degenerated lands dominated by poverty, in India, these became economic hubs, which are informal, but they work (Tiwari, 2014).

The living condition in cities has become so poor due to crumbling services from local authorities that it has caused disenchantment with the political system. In a country which fought for independence for democratic governance through mass movements, people have started to disengage with the local political system. Another kind of fortification has emerged where the rich live in gated communities where services are privatised and security is privatised. The market-based development agenda is leading to an unplanned growth on urban fringes, as discussed in Chapters 2 and 5.

The urban form that has been added since independence, as discussed in Chapter 2, is a contrasting reflection of inequalities at many levels.

1.6 The structure of the book

The foregoing discussion forms the contextual environment with which cities of Delhi and its surrounds have evolved. The built form that we see today is a reflection of politics, policies, economic and social endeavour of humanity who has lived here. The rest of the chapters explore a range of questions about urban form, the material properties, design and sustainability that have been underpinned by the legal, social and economic systems that influence and shape urban form, and rights, claims and conflicts over access and use of urban resources. These chapters present an inquiry into the continuities or discontinuities in urban form, architectural design, construction materials, land and economic policies, regulation and institutions and how these have shaped Delhi and its surrounds that we see today. Set out next is a summary of these chapters.

Chapter 2: The market, policy and regulatory context of Delhi and its surrounds

This chapter investigates the role of market, finance and economy and policies around these in shaping the built environment. In the context of Delhi and its surrounds, this chapter examines the extent and effectiveness of policies and instruments and institutions for effective policy implementation to address the challenges that the built space has faced. The chapter highlights the visible impacts of lack of responsive planning and infeasible regulations on the city's physical form and natural environment and economic growth pattern.

Chapter 3: Evolution of urban form and imageability

Starting with what is 'urban' and explanations around their formation and expansion, the chapter looks deeper into the history of evolution of the urban form of Delhi while analysing the drivers which were influencing the then city form and pattern of growth. Contemporary cities of various times and eras have been studied alongside the then Delhi. All stories are knit together to create an overall picture of the social, political and economic situation which was prevailing and was guiding the form and scale of the built environment of India in general and Delhi and its surrounds, in particular.

Chapter 4: Continuities and discontinuities

Taking forward the discussion of 'imageability', this chapter carries forward and converges the discussion into the physical outlook, design and materiality of buildings or monuments constituting the physical space of the city of Delhi while responding, or conflicting, with the climate and geography of the place. Through an investigation of historical evolution of Delhi and contemporary influences of globalisation, this chapter explores both natural and artificial materials, local and foreign designs and materials that the city has used in its construction. This chapter presents discourse on continuity/discontinuity in architectural designs and materials over time.

After having known the physical form, its components and its constituents, it is interesting to analyse how city's physical character has contributed or conflicted with its sustainability; how, with the passage of time, the responsiveness towards the environment has been reducing, and the mention of the issue in policies and regulations has been increasing without much implementation. Using examples from architectural design and construction of elements from various times, this chapter also investigates how the city responded towards sustaining resources like water. The discussions include arguments and counterarguments over drawing a balance between economic development and the environment. The chapter initiates a discussion about economic growth, the increasing demand for functional building space and associated issues and claims over land and property.

Chapter 5: The intersection of surrealism, welfarism and consumerism

The 'whole-to-part' approach of discussion about Delhi starting with description of evolution of urban form and concluding with the interweaving of macro issues with city development decisions, both in the formal and in the informal domains, will be reversely reviewed with 'part-to-whole' approach so as to recapture the outward effect that the contemporary development of city is having on the overall shape and form of the city.

Notes

1 A brief mention of Indus Valley is considered important because of it being considered a forerunner of the 'second urbanisation' of India along the River Ganges, of which Delhi is a part. However, this book will not cover the details of the Indus Valley Civilisation, which is otherwise very well studied and documented.
2 Basham (1969).
3 A watershed is an elevated area which keeps two river systems distinct and separate from each other (Thapar, 2002).
4 Mahabharata is one of the two most important epics of ancient India. There is no surety of the date to which the descriptions of the two epics could be assigned and the dating of the epic is itself doubtful (Thapar, 2002). Ray (1964) writes that "[i]n the form they are known to us, the Ramayana cannot be dated earlier than the 2nd century A.D., and the Mahabharata not earlier than the 4th century" (p. 54). Thapar (2002) writes that "the core text (of Mahabharata) . . . is generally believed to have existed by about the middle of the first millennium B.C. or possibly earlier" (p. 91). However, there is no archaeological culture or site that represents the descriptions in these epics (Ratnagar, 2002).
5 Mahabharata (1.19), referred by Thapar (2002, p. 98).
6 Basham (1969).
7 Refer Wagle (1966) for more details on administrative units like *gram*, *nagar*, *janpada*, *mahajanpada*.
8 Basham (1969).

Bibliography

Allchin, B., & Allchin, R. (1968). *The Birth of Indian Civilisation*. Great Britain: Richard Clay (The Chaucer Press) Ltd.

Bartuska, T. (2007). The Built Environment: Definition and Scope. In W. B. McClure (Ed.), *The Built Environment: A Collaborative Enquiry into Design and Planning*. Wiley.

Basham, A. L. (1969). *The Wonder That Was India: A Survey of the History and Culture of the Indian Sub-Continent Before the Coming of the Muslims*. London: Sidgwick & Jackson.

Blake, S. (1991). *Shahjahanabad: The Sovereign City in Mughal India*. Cambridge, UK: Cambridge University Press.

Buckler, F. (1922). The Political Theory of Indian Mutiny. *Transactions of the Royal Historical Society*, 71–100.

Centre for Iranian Studies. (2017). *Delhi Sultanate*. Retrieved from Encyclopaedia Iranica: www.iranicaonline.org/articles/delhi-sultanate

Chattopadhyay, B. D. (2012). *The Making of Early Medieval India*. Oxford University Press. Delhi.

Childe, V. G. (1950, April). The Urban Revolution. *The Town Planning Review, 21*(1), 3–17.
Editors of Encyclopaedia Britannica. (2017). *Mughal Dynasty*. Retrieved June 1, 2017, from Britannica Encyclopaedia: www.britannica.com/topic/Mughal-dynasty
Fletcher, R. (1995). *The Limits of Settlement Growth*. Cambridge: Cambridge University Press.
Harrington, A. (2005). Introduction: What Is Social Theory? In A. Harrington (Ed.), *Modern Social Theory: An Introduction*. Oxford: Oxford University Press.
Hermann, K. & Ruthermund, D. (2004). *A History of India*. London: Routledge.
Hobbes, T. (2010). *Leviathan: Or, The Matter, Forme and Power of a Common Wealth Ecclesiasticall and Civil*. New Haven, CT: Yale University Press.
Ifthikar, R. (2017). *Working Class in Mughal India (1556–1605)*. Retrieved June 1, 2017, from Punjab University: http://pu.edu.pk/images/journal/history/PDF-FILES/10-%20Rukhsana%20Iftikhar_52-1-15.pdf
Ikram, S. (1964). *Muslim Civilisation in India*. New York, NY: Columbia University Press.
Jackson, P. (2003). *The Delhi Sultanate: A Political and Military History*. Cambridge: Cambridge University Press.
Keogh, G. & D'Arcy, E. (1994). Market Maturity and Property Market Behaviour: A European Comparison of Mature and Emergent Markets. *Journal of Property Research, 11*, 215–235.
Khosla, R. & Rai, N. (2005). *The Idea of Delhi*. Mumbai: Marg Publications.
Kosambi, D. D. (1956). *An Introduction to the Study of Indian History*. Bombay: Popular Book Depot.
Kosambi, D. D. (1972). *The Culture and Civilisation of Ancient India in Historical Outline*. New Delhi: Vikas Publishing House.
Kulke, M. & Rothermund, D. (1999). *A History of India*. 3rd Edition, London: Routledge, p. 103.
Lal, K. (1967). *History of the Khaljis*. Asian Publishing House.
Lal, M. (1984). *Settlement History and Rise of Civilization in Ganga-Yamuna Doab, from 1500 B.C. to 300 A.D.* Delhi: B.R. Pub. Corp.
Nehru, J. (1981). *The Discovery of India*. New Delhi: Oxford University Press.
Paul, J. J. (1996). Historical Settings. In J. Heitzman & R. L. Worden (Eds.), *India: A Country Study* (pp. 1–60). Washington, DC: Federal Research Division, Library of Congress. Retrieved April 20, 2017, from www.loc.gov/item/96019266/
Ratnagar, S. (2002). Archaeological Perspectives on Early Indian Societies. In R. Thapar (Ed.), *Recent Perspectives of Early Indian History* (pp. 1–59). Mumbai: Popular Prakashan Pvt. Ltd.
Ray, A. (1964). *Villages, Towns and Secular Buildings in Ancient India: c. 150 B.C.–c. 350 A.D.* Calcutta: K.L. Mukhopadhyay.
Raychaudhuri, T., Habib, I & Kumar, D. (1982). *The Cambridge Economic History of India: Vol. 1*. Cambridge: Cambridge University Press.
Scaff, L. (1995). Social Theory, Rationalism and the Architecture of the City: Fin-de-Siecle Thematics. *Theory, Culture and Society, 12*, 63–85.
Sharma, R. S. (2005). *India's Ancient Past*. Oxford; New York, NY: Oxford University Press.
Sharma, R. S. (2006). *Feudal Polity in Three Kingdoms in India, in, Indian Feudalism c. AD 300–1200*. Kolkata: University of Calcutta.
Sinopoli, C. (1994). Monumentality and Mobility in Mughal Capitals. *Asian Perspectives, 33*(2), 293–308.

Spear, P. (1951). *Twilight of the Mughals*. Cambridge: Cambridge University Press.
Squires, G. & Heurkens, E. (2015). *International Approaches to Property Development*. London: Routledge.
Thapar, R. (1987). *Ancient Indian Social History: Some Interpretations*. New Delhi: Orient Longman Limited.
Thapar, R. (2002). The First Millennium B.C. in Northern India (Up to the End of the Mauryan Period). In R. Thapar (Ed.), *Recent Perspectives of Early Indian History*. Mumbai: Popular Prakashan.
Tiwari, P. (2014). Crush Hour: Explosive Urbanisation and Its Consequences for India (L. Haultain, Interviewer). *Up Close*.
Wagle, N. (1966). *Society at the Time of the Buddha*. Bombay: Popular Prakashan.

2 The market, policy and regulatory context of Delhi and its surrounds

Review and assessment

2.1 Introduction

The role of market, finance and economy, and the policies around these that impinge on land and shape the built environment cannot be underestimated. The extent and effectiveness of policies and institutions have significantly impacted Delhi's physical form and natural environment. In this chapter, a brief historical overview of market, policy and regulatory context as they relate to land and the built environment is presented. The chapter also discusses the land use and planning approaches that Delhi and its surrounds used in its development post-independence. It may be highlighted here that the discussions on market, policy and regulations that affected Delhi have often got intermingled in the discussion in this chapter. A discerning reader may find a very selective discussion of these rather than a comprehensive analysis. Partly this is due to the overall objective of the book, which is to understand built environment for which market, policy and regulations provide an overarching institutional framework, as discussed in Chapter 1, and partly because of the limitation of research in this area that has analysed the impact of these on the built environment. Despite these shortcomings, this chapter provides an important context to see the dynamic evolution of space in Delhi and its surrounds.

The seven cities of Delhi as described in Chapter 1 are largely the construct of 700 years of Sultanate and Mughal rule. The eighth city, New Delhi, was constructed during the British time, which has retained its status as the capital of post-independence India.

The chapter is organised as follows: Section 2.2 discusses the land rights during the Mughal rule and explores the notion and extent of freehold. In Section 2.3, land rights, which underwent phenomenal changes during the administration of the British East India Company, are discussed. Section 2.4 discusses the making of New Delhi and the role of regulations such as the Land Acquisition Act 1894. Section 2.5 discusses the shift in development paradigm after independence. Section 2.6 presents a brief overview of institutional development related to the built environment in Delhi during the pre-Master Plan era of 1947–57. Section 2.7 reviews the impact of Delhi Development Authority, a planning and development agency that was established to master plan Delhi's growth and also undertakes development activities to meet housing needs of the city. Growth on urban fringes of Delhi led to the formation of National Capital Region and National Capital Region Planning Board, as discussed in Section 2.8. The current challenges, in particular related to housing, in Delhi and its surrounds are discussed in Section 2.9. The impacts of these policies and programmes on built space on the urban fringes of Delhi are discussed in Section 2.10. Finally, Section 2.11 concludes.

2.2 The treatment of land during the Mughal period

The record on land and its attachment during the Mughal period has to be derived implicitly from administrative 'land revenue' arrangements of the time. The modern form of private 'freehold' ownership was an alien concept during the Mughal period, and the idea of land ownership in Western sense was unknown (Pearson, 1985). Based on the records of European travellers during the sixteenth and seventeenth centuries, as discussed by Raychaudhuri et al. (1982), land ownership was vested in the emperor, due to which the emperor had the right to appropriate rent from the produce of the soil. Pearson (1985) clarifies that the land revenue was "not a rent paid by a landholder because he was using royal property" and that he paid it as "remuneration of sovereignty" for "the protection and justice provided".

Mackenzie, as cited in Raychaudhuri et al. (1982), presents the view of British administrators later that 'land revenue' was considered as 'property' (granted to king through long history) as it represented almost ten-elevenths of the net rental in the country. Some, such as Muhammad (cited in Raychaudhuri et al., 1982), argue that this was not the case and the tax levied during the Mughal time was tax on the crops and not on land. Pearson (1985) argues that Mughals wanted control, not ownership, and for that they needed to assert power to control resources such as land and the people on it. Their main thrust of rule was to extend their power, reduce the influence of intermediaries and extract as much surplus as possible for themselves (ibid).

In the early period, the tax was a share of the actual crop output, which required an extensive burden on the administration to collect, and later of an estimated output based on a yield-based formula to ease the administrative burden. During Sher Shah Sur's rule (1540–45 AD), the tax basis was shifted to the area of land sown rather than actual harvest. This likely recognizes land as 'property' of the emperor. This system was followed in Delhi and possibly in the rest of the country under Sher Shah Sur's rule (Raychaudhuri et al., 1982). Later, during Akbar's reign (1574–75AD), the tax system shifted from a system of rate based on crop to cash. The cash rates, adjusted for type of crops, were applied to land area under cultivation. The data on crop yield, prices and area were collected to determine the highest and best use of land, which then was used to set the cash rate. The system was quite elaborate, and provinces including Delhi were divided into revenue circles.

While Mughal rulers desired to assess the tax burden at the cultivator level based on the size of land cultivated, it was practically impossible to administratively manage collection of taxes at that level of disaggregation. This required them to establish intermediaries (zamindars) to act on their behalf. The revenue units were villages and assesses were zamindars, who often had bought these rights from emperor and settled peasants to cultivate large land, which were cleared for agriculture. Zamindari rights were an article of property and could be inherited, monetised, mortgaged, sold or confiscated by ruling class but did not amount to absolute ownership of land. In theory, claim to land revenue and other taxes belonged to emperors, but temporary alienation of specific areas in favour of members of a small ruling class (*Jagirdars*), who held ranks granted by the emperor, was practiced. These land grants were temporary and short term and did not remain fixed for the grantee (Raychaudhuri et al., 1982). Very few rank holders were local zamindars, and a large proportion of them were immigrants (Turanis, Iranis and Afghans). Bureaucrats and intelligentsia also constituted small number of beneficiary of this system. The overall implication was that the land had got concentrated in the hands of a few. However, the dependency of these grant holders on the emperor was paramount, as these grants were neither geographically permanent nor inheritable (Raychaudhuri et al., 1982).

Land rights were unbundled as reflected in the power relations between different classes with emperor as the absolute owner who had total right in land revenue, which he claimed through a system of zamindars. For Mughals, the area of prime concern was Doab and Subas of Delhi, Agra, Oudh and Allahabad. Here the land control was directly connected with their power and prestige. Land was either crown land or granted to princes (Pearson, 1985). For other locations, land was temporarily granted to *Jagirdars*, who became the claimants of land revenue from zamindars, albeit for the period of grant. The ownership of zamindars on land is reflected by an observation made by Raychaudhuri et al. (1982), that "so great was the *Jagirdars*' power that . . . the hakim (*Jagirdar*) of the day can in a moment remove Zamindar of five hundred years and put in his stead a man who has been without a place for lifetime". *Jagirdars* claimed power to detain cultivators on land like serfs and could bring them back if they ran away (Raychaudhuri et al., 1982). There were also a very small number of land grant holders who held these for their lifetime, and upon their death, the land grant was conferred on their offspring. These were Muslim scholars, pensioned-off government officials, widows and other women belonging to bureaucratic social class. Due to the relative permanency of these land grants, some grant holders sought to acquire the rights of zamindars on their land grants or elsewhere.

The village community was divided into stratum of cultivators, from landless labourers at the lowest end of the ladder to peasants to large cultivators, who had different levels of rights on the produce. Zamindars held rights over few villages and *Jagirdars* over much larger area. Other systems, such as peasant villages with no recognisable zamindars, also existed. The emperor could establish new zamindaris, and expand or abolish existing zamindaris. The agrarian relation indicates a economic structure where, as observed by Bernier (cited in Raychaudhuri et al., 1982), a large portion of peasants were poor, and their conditions were appalling and wealth was more and more concentrated with each upper layer of society. Taxation was regressive and the system oppressive.

The interesting corollary to the preceding discussion for built space is that the houses of peasants were very basic hutments. Zamindars had forts as they had to maintain administrative organ, which at times included small armies, to collect revenues from peasants. *Jagirdars* had much larger forts and much larger armies under their control. Given that much of the agricultural surplus was collected as revenue, there were enough resources left with the ruling class to build grandeur buildings and maintain an opulent lifestyle.

Urban areas were largely producers of manufacturing goods. There are conflicting views among historians whether India was a leading manufacturing nation exporting fine manufactures at least at par with leading pre-industrial Europe or technologically primitive, stagnant, low productivity manufacturing centre. Lack of data and historical accounts inhibit conclusive answers, but the trend in "manufacturing goods was almost certainly upwards" (Raychaudhuri et al., 1982). The surplus that rural areas produced was captured as land revenue, and concentration of this wealth in few hands generated demand for manufactured goods in cities. In addition, a sizeable middle-income class was formed, comprising lower-rank bureaucrats, professionals and holders of rent-free tenures who lived in cities and demanded comfort and luxury goods.

There are historical records of that period that list a range of craft and service providers that developed in cities (Raychaudhuri et al., 1982), which implies that a thriving centralised market place comprising artisans' workshops and shops would have formed an important part of cities. Manucci (quoted in Raychaudhuri et al., 1982) describes Delhi as a place "where everything finds a sale and are consumed". Shahjahanabad, as administrative capital of the Mughal Empire since 1639, was a major importer of goods manufactured by artisans in other cities. The biggest consumers of these goods were the household of the emperor and the nobility.

Rich traders and growing middle-income groups demanded better and bigger houses. In Delhi, richer merchants lived like nobles in imposing buildings that were at times three stories in height. The lower bureaucracies also had decent buildings, which appeared reasonably good and well ventilated. All seven cities of Delhi were walled cities, but the growing population and urban sprawl always caused the development to leap over the walls. Shahjahanabad remained the capital of Mughal India until 1857.

2.3 Land tenure during the administration of the British East India Company

The most important shift in the land tenure system in India occurred when the British East India Company entered into an agreement (permanent settlement) with the zamindars (landlords) in Bengal in 1793. The idea of permanent settlement drew on prevailing economic thoughts in France and the UK particularly around "free trade and the perceived social and economic benefits brought by securing individual property ownership" (Guha, 1996). The British East India Company administration experimented with three major land assessments and settlement formats in India, which were zamindari[1] in Bengal, as discussed earlier, *Raiyatwari*[2] in Madras (now Chennai) and Bombay, and *Mahalwari*[3] in the North-Western Provinces[4] and British Punjab (Husain & Sarwar, 2012). While zamindari system gave immense powers to the zamindar who exercised control over land, under other two systems the peasant had ownership rights and was also liable for taxes.

Husain and Sarwar (2012) discusses geographical coverage of the three systems: the permanent zamindari settlement was covering 19% of total area of British India; the *Mahalwari* system covered nearly 30%; and the *Raiyatwari* system covered roughly 51% of the total British Indian territory.

- Permanent zamindari settlements were made in Bengal, Bihar, Orissa (now Odisha) and the Banaras divisions of Uttar Pradesh This settlement was further extended in 1800 to Northern Carnatic (northeastern part of Madras) and North-Western Provinces (eastern Uttar Pradesh).
- The *Mahalwari* tenure was introduced in major portions of Uttar Pradesh, the Central Provinces, the Punjab (with variations) and the central providences; while in Oudh, villages were placed under taluqdar or middlemen with whom the government dealt directly.
- The *Raiyatwari* settlements were made in major portions of Bombay, Madras and Sindh Provinces. The principles of this system also applied to Assam and Burma. A few hilly tracts in Bengal and the coast strip of Orissa had been temporally settled.

Johnson (2015) argues that while in theory, individual ownership of land and its productive potential that permanent settlement ensured would create a stable, satisfied and wealthy rural India, in practice, individual landowners were saddled with huge debt from local money lenders to settle the land revenue obligations and to buy necessary inputs for agriculture. The creation of free market in rural India led to concentration of land in the hands of those who had investible capital, whether they were agriculturists or not and in many instances cultivators became tenant farmers. The resultant outcome resembled feudal system, precisely like the one the introducers of these land reforms were intending to avoid. The permanent settlement laid the foundation of how the British managed the land tenures and land revenues in India (ibid).

These three systems of land tenures and land revenue arrangements underwent lots of changes during the British Company administration, and this resulted in the intermixing of features, which ultimately tended towards the zamindari system so as to serve the purpose of intensive revenue generation the best, with a single point of contact as the 'zamindar'. The hierarchy of the agrarian system was complicated, as there were several layers operating between the zamindar and actual tiller. Zamindars were allocating *pattas* (lease) to the cultivators for rent failing which the zamindars were authorised to use their right of 'tenant eviction', buying and selling of land; and had the "freedom to use physical coercion on their own account, without the permission of a court" (Swamy & College, 2010). From rent received by zamindars, they had to give a previously agreed amount to the British East India Company 'collector',[5] failing which the company had the authority to take over all land holdings of the zamindars.

Delhi and its surrounds also had the characteristic land tenures, which reflected *Mahalwari* and zamindari systems. With the concentration of land through sale by cultivators to large landowners or money lenders, the system moved move close to the zamindari, though it is plausible that *Mahalwari* was the dominant system.

Geographically, Shahjahanabad was a large city between a rocky ridge on the west and the River Yamuna on the east. To the north of the city, the area between the river and the ridge became narrower, and the land opened in a marshy land. European Civil Lines and cantonments were located between Shahjahanabad and the marshy land. Shahjahanabad was surrounded by hundreds of small agricultural villages, typically of an average area of around 1,100 acres with about 900 inhabitants. The southern part of the city was largely agricultural, with intermitant Hindu and Muslim settlements established during different political regimes that existed prior to Shahjahanabad. Firozabad served as the suburb of Delhi. The region was arid, and the land was characterised by low agricultural yields. Though two-thirds of the land were under agriculture, the agricultural output was low. Canals and wells irrigated half of the agricultural area under cultivation, and the rest was dependent on monsoons. Low productivity meant that a large population was poor and there were few rich and powerful landlords. The average landholdings of 2.2 acres were small, and from historical land records it was difficult to distinguish between peasent and the zamindar, given that both had comparable sizes of land holdings. In addition, there were seasonal labourers who professed other trades, such as leather-makers, washermen, sweepers etc. who worked on the land during harvest periods. These rural communities were bound by ties of reciprocity between individuals and loosely held together by vague sense of land rights (Johnson, 2015).

2.4 Land tenure during the British Empire

Following the defeat and exile of Mughal emperor in 1857, the foundation for the British Empire in India was laid, and the rule of the British East India Company was transferred to the British Crown on 28 June 1858. As a new colony of the British Empire, development works required land. This necessitated an instrument to expropriate land to carry out development works of the British Empire in its new colony. The Land Acquisition Act 1894 was passed which facilitated the acquisition of private land for public purposes. Previous systems of the British East India Company, which focused narrowly on land revenue, had dispersed land ownership. The Registration Act of 1908 required that the "documents" relating to real property rights are registered in the office of Sub-Registrar within whose jurisdiction whole or some portion of property was situated. If these documents were not registered under the Act, these could not be subsequently introduced in the court of law as claim to property right (Tiwari et al., 2015). The Act ensured that the land records were very well maintained up to the extent of allocation of distinct Khasra[6] number to agricultural land holdings serving the British Empire's interest in maintaining a good system for collection of land revenue. As observed by Panagariya (2008), the records contained details on cultivable land area, soil quality, source of irrigation, cropping patterns – but did not include details about the ownership of land. Owner details were not important, especially when it involved complications and could have invited disputes.

Revenue settlement was last carried out in Delhi in the years 1908–09. The lands earmarked for village settlement (*abadi*) and those meant basically for agricultural purposes were duly demarcated. The village *abadi*, that is, essentially the residential component of the community, was shown in the village map circumscribed in red ink. The *abadi* area came to be known as Lal Dora in common parlance. Lands falling within Lal Dora were not assessed to land revenue. The agricultural fields outside the *abadi* were subject to land revenue (MoUD, 2007). During the process, the *abadi* was recorded under one Khasra number without any further detail on ownership as the same was not assessed to land revenue and was therefore less important to be documented in detail. The purpose behind maintaining land records was to facilitate tax collection and not to determine the title of the land. Lack of record of ownership with Lal Dora areas rendered properties within Lal Dora to remain mostly unmapped and unregistered and the occupation of these properties ran on family lines (Pati, 2015).

With the announcement of the shift of capital of the British Empire in India from Kolkata to Delhi in 1911 by King George V, a large-scale acquisition of land to build New Delhi followed. New Delhi was a project of British imperial aspirations to reinterpret the meaning of British Empire in India during the twentieth century (Johnson, 2015). Acquisition of land to build New Delhi followed land policies long used in British India. Britain's political economy in India that was focused on land revenues derived from privately held lands had changed little when the decision to build the new capital was made (ibid). The most important among these was the Land Acquisition Act, 1894. In its firm commitment to free market economy, which had shaped the land laws, the British Empire applied the Land Acquisition Act 1894 indiscriminately to acquire land. This rendered many farmers landless in the process. Britain's reluctance to forego free market principles even during the time of droughts, which had occurred in the periods preceding the construction of New Delhi, had left many farmers in financially dire conditions. Under the Land Acquisition Act, fair compensation and resettlement were offered to those who had titles to the land, but no allowance was made for the rights of others without titles in land proceedings. Many landless rural residents who worked on the lands of those whose lands were acquired were set adrift. British land policies privileged private land ownership and free markets over community (Johnson, 2015), as many displaced landowners were resettled in more fertile, canal-irrigated lands in Punjab.

The new capital, built between 1912 and 1931 by Edwin Lutyens and Herbert Baker, was an ambitious project reflecting the assimilation of India into the British Empire bound by a vision that aimed to give more political freedom to Indians while simultaneously binding them to the British Empire (Johnson, 2015). The new capital was hence to be built with a mixture of Indian vernacular and Western architectural styles. British India's colonial building traditions in India reached the greatest synthesis in New Delhi, demonstrating that the strength of British Empire lay in colonising Indian knowledge (ibid). In a major deviation from the British buildings in Calcutta, in building New Delhi, British town planners and architects attempted to incorporate authentic Indian architectural styles that would blend with Western styles, reflecting Hardinge's argument during planning stages that the city must reflect "a broad classical style with an Indian motif" (ibid). In a letter to Baker, Hardinge wrote (as cited in Johnson, 2015) "the aim must be to achieve a style which will be symbolic of India of twentieth century, with its British and Indian administration". The buildings must then appeal to Indians from their Indian sentiments (ibid). The attitudinal change in British colonial rule was a response to growing influence of nationalist movement. Growing demand for independence and increasing passive resistance to British rule led to change in political approach that involved more and more Indians in administration while retaining the overarching authority with the British. With the Government of India Act 1935, the British offered a colonial federalism so the provinces in British India were administered by elected Indian officials, while the Britain maintained ultimate authority at the centre in New Delhi.

The new capital required 40,000 square acres of land, a large proportion of which was acquired by force from poor Indian cultivators. The land officers had to balance two competing objectives in using the Land Acquisition Act of 1894: to adhere to the law and to keep the cost of acquisition low. The Land Acquisition Act allowed private property to be reapportioned to the public sector. The law allowed a compensation based on the fair value of land to be paid to the dispossessed, and the compensation was based on the market value as on date when the gazette notification to acquire land was published. The low-producing agricultural land of Delhi, which was mired with droughts in years prior to acquisition, fulfilled the criteria of low-cost acquisition. A new administrative set-up comprising officials from district's land revenue office including British and Indian officials carried the acquisitions. Immediate and future needs for land were assessed, and villages for acquisition were measured in detail. Addison (cited in Johnson, 2015) in his 'Report on Land Acquisitions in Delhi, 1915' reports that the total cost of acquisition was Rupees 4.82 million. Three methods of fair compensation payment for acquired land were used – cash payment, revenue remissions and alternative land. It was argued that large landowners could be paid in cash. Alternative land was a preferred method for small landowners, but the difficulty in obtaining comparable (of similar value) arable land around Delhi became a challenge; only a third of required land to pay as compensation could be procured. Land in the Lower Doab Bari Canal area was expensive compared to Delhi.

There are evidences of innovative structures of compensation payment. As Johnson (2015) explains, in one case in lieu of 0.25 acre of land lost in Delhi, one farmer was compensated with a high-value standard size of land plot (25 acres) in Lower Doab Bari Canal area. However, an agreement was made wherein the land would meet the immediate needs of the farmer and the farmer would pay for the excess area by future instalments. Large landowners whose lands were acquired chose to settle in Lower Doab Bari area, and they used their cash to pay for additional area beyond what compensation for their acquired land would have allowed. For smaller farmers the compensation was meagre. They did not have enough resources to finance their move from Delhi district to Lower Doab Bari area, let alone have resources to farm there. There were also concerns about breaking of social ties, the loss of which was not accounted for in the Land Acquisition Act. The compensation was viewed purely from market economic point of view in the law. Not more than 20% of displaced landowners accepted compensation in the form of land in the Lower Doab Bari Canal area (Johnson, 2015). There were many landlords who benefitted from the acquisition, as they could negotiate better compensation, lands in Lower Doab Bari Canal area far greater than what they lost in Delhi. Some through their proximity to the British (like the Sikhs in the Rakabganj area) were not only able to get better compensation but also could protect their *gurudwara* (religious place for Sikhs) from getting demolished. The losers in the land acquisition process were those who worked on the land but did not have land ownership as the Act compensated only the landowners. These peasants were set adrift. They had lost their ties with the previous landowners in whose fields they worked as the landowners moved.

With the acquisition of land, the Delhi district lost its source of land revenue. The British government of India encouraged the land acquisition officers to work out an arrangement with zamindars and peasants to continue cultivating the acquired land on payment of a rent till it was finally required for construction. Many landlords and peasants who did not want to move continued to stay and cultivate and pay land revenue albeit as tenants of the British government of India. This allowed landlords and peasants to maintain their way of life, and some continued to live there as long as till 1920s. The government also benefitted from this system, as the source of revenue continued which even funded the initial land acquisition.

Given that this chapter aims to discuss the markets and policies within which Delhi evolved, it would not be inappropriate to state that the market economy and the policies to support that, which the British Indian government pursued, suited the financial interests of the British Empire rather than of the native Indians. British government of India became the largest landowner (similar in some ways to what the Mughals were during pre-British occupation times), and the impoverishment of the bottom of Indian society continued. The resistance to land acquisition was less, as there were droughts and the landowners were in a dire economic state. The high cost of legal processes also precluded them from resisting. Moreover, fragmented land ownership had meant that the opportunity for large organised resistance against land acquisition was limited.

There were, however, a number of factors that were causing disruptions to the urban fabric of post-Mughal period since 1857. New technologies such as railways (1867), waterworks (1892), electric trams (1901–02), electricity (1902), drainage (1909) and telephones (1923) were introduced in the old city through colonial interventions, and the area grew as a commercial hub (Mehra, 2013). Buildings within the city walls were expanded and subdivided to house commercial activities and new residents as the population increased (ibid). The city boundary had already extended beyond the wall, and it started growing in the west and north as shops, factories and mills were established. The city was becoming congested and unhealthy. The Delhi Municipal Committee (DMC) was inadequately equipped to deal with the situation, as it was cash strapped and lacked technical expertise. In contrast, the new imperial capital, New Delhi, was a sprawling extravaganza with spectacular buildings, big boulevards, wide roads conducive for cars and large mansions with gardens. The contrast was obvious as the Old Delhi housed the native population, while the New Delhi became home to the 'British Government of India' officials, most of them non-native British.

Under public pressure to address appalling living conditions outside New Delhi, a new institution, the Delhi Improvement Trust (DIT), was established. The mandate of DIT covered both the New Delhi and DMC areas. The DIT was tasked with building houses to meet the burgeoning shortage, and for this, the agency requested that they be allowed to use government land that was acquired either through the land acquisition process and was lying unused or the land that was allocated by the Mughal emperor around Old Delhi. The government of India agreed to these but required the DIT to pay an annual rent for these lands and to return any profits. The cost of town expansion and provisioning of services was to be financed using the profitable sale of leases of the DIT's lands after their development. With these requirements, only developing new middle- and lower-middle-class housing was found to be financially viable. Where applicable, the cost of housing for the poor was to be offset by profits made by selling upper-class housing (Mehra, 2013). For better control on the buildings on government-owned 'surplus' lands, DIT started leasing plots using a 90-year lease. This allowed the DIT to secure any increment in land values that was due to improvements effected by it. After 1941, the DIT Trust would sell plots on perpetual leases to the highest bidder by auction or by tender. The buildings that were approved strictly followed the building by-laws (Mehra, 2013).

India got drawn into World War II as Allied forces used India as a base to resist advancing Japanese forces. The open areas of New Delhi became the camping grounds for Allied forces. Cheap, modular, temporary, semi-permanent accommodation for the forces and war bureaucracy was constructed. The population of Delhi increased manifold. The heightened industrial activity to meet the needs of war personnel led to the in-migration of workers from other parts of the country. These conditions made the housing situation much more acute, and with ensuing of war, DIT had not been able to contribute much in terms of new housing even though the land had been procured. The Rent Control Order of 1939 (replaced by an ordinance in 1944) and the Punjab Urban Rent Restriction

Act 1941 capped an increase in rent in New Delhi and DMC areas during the war. New housing could not be easily built due to the shortage of building materials (Mehra, 2013). All this contributed adversely to the already-precarious housing situation.

During the war, as Mehra (2013) discusses, DIT sought to benefit from rising land prices by raising ground rents or selling plots at higher prices through auction to war-rich. Its primary objective to meet the housing shortage could not be served. Most wartime controls continued even after the war, as the British Indian government was under pressure on many counts – the mounting financial burden, the growing nationalist movement for independence and the collapsing empire.

2.5 Post-independence shift in the development paradigm

2.5.1 *A developing state*

Absence of a welfare state and perusal of a market economy during the rule of the British Empire meant that when India got independence, two in three Indians lived in absolute poverty. The British attempt to modernise the economy through building of the railways, canals, establishment of property rights and commercial law, textile mills in Mumbai and Ahmedabad and iron and steel industry in Bihar and Orissa instilled the principles of market economy; however, it left the economy in a dire state with the weakening of local craft industry and agriculture (Tiwari, et al., 2015).

New development paradigm that marked a distinct departure from British market philosophy required that the social and economic modernisation of India would have to be achieved through concerted planned actions emanating from New Delhi. However, the initial conditions for building development paradigm were very weak. The nation inherited a splintered society – fractured by caste, class, economic disparities, rural-urban divides and a multitude of beliefs and religious affiliations that were wielded together as a nation-state. Modernisation of industrial practices had not happened, and the productivity of agriculture sector was very low. The partitioning of the country and the largest human migration in the history of humanity that followed left a huge social and economic burden that the nation had to shoulder in years to come.

The economic thoughts of the nineteenth century influenced the British approach to development in India. Ricardo, in 1817, in his treatise on trade argued that countries should specialise in production of those commodities in which they have comparative advantage and engage in trade by exporting the same and importing those for which they don't have comparative advantage. India became an exporter of raw material and an importer of manufactured goods from Britain. Moreover, since the prices of manufactured goods rose faster than those of the primary goods, while the trading partners gained from such a trade, it weakened the Indian economy and also negatively impacted local manufacturing which was decentralised using traditional tools and equipment.

Consequently, the post-independence economic paradigm was based on self-reliance, protectionism and directed growth emanating from a strong centre with the hope that this top-down approach would percolate to states and from states to towns and villages. Self-reliance implied import-substitution industrialisation, which required the production of capital goods such as iron and steel, chemicals, heavy engineering (Tiwari et al., 2015). Strict controls on foreign exchange to promote import-substitution industrialisation and poorly formed capital markets inclined the government of India to think "economic development as a project that had to be planned and delivered by a beneficial state" (Corbridge, 2009, p. 6). India was to be governed by a development state model with a Planning Commission and its Five-Year Plans playing a central role. It was assumed that the government of India would be able to funnel resources from agricultural sector to non-agricultural sector without much rural backlash (ibid). In a major departure from British India's economic model, the first wave of capital-goods-based production was not to be labour-intensive. Once self-reliance on capital goods was achieved, these then would be combined with labour from the countryside to produce consumer goods, and the Arthur Lewis (1954) model of economic development would be realised in the second stage of industrialisation (Tiwari et al., 2015).

But this was too presumptuous, and a mass farmers' leader, Charan Singh, in his writings on India's agriculture and agriculture policy, anticipated that India was suffering from an "urban bias" (Corbridge, 2009). Opponents were quick to point out that in a country where three-quarters of the population lived in rural areas and agriculture produced more than half of GDP, it made little sense to waste capital on inefficient urban and industrial projects. Instead the effort should be to fund irrigation infrastructure and create off-farm employment in rural areas (Tiwari, et al., 2015). The view gained further currency as the country faced failure of monsoon in 1965 and 1966, and the new data showed a rise in rural poverty. The economic situation of the country was further tattered after a disastrous war with China in 1962. All these conditions and the deepening democracy changed the course from "command politics" to "demand politics". There was a rise of credible opposition parties, and this was also a time when a "prospectively developmental state imploded" (Corbridge, 2009).

The modernising agenda of Nehru driven by the progressive elite had to negotiate at the local level with politicians who rarely shared the same commitments to larger benefit or the long run. India's modernist state "had feet in vernacular clay" (Kaviraj, 1984). As discussed by Corbridge (2009), the development state was captured by three interlocking groups: richer farmers (who blocked agrarian reforms), industrial bourgeoisie (who took advantage of state-induced scarcities and which blocked competition and innovation) and the country's leading bureaucrats (who benefitted from Permit-License-Quota-Raj) (Bhagwati, 1993). The state was forced to accommodate to the demands of various interest groups irrespective of whether they were in the larger interest of state or not. Consequently, the economic growth in the 1970s suffered.

At the systemic level, the failure of the Congress Party in the 1950s and 1960s to support a development state led the economy to fall between two stools. One, state was not strong enough to force economic growth with sound fiscal structure. The state could not do away with subsidies and protectionist barriers that were meant to be temporary (Bhagwati, 1993). Second, the central role occupied by the state in India's productive economy stifled innovation and new start-ups in the organised private sector. The political platform of Congress in the 1970s became "garibi hatao (eliminate poverty)", but this was never achieved, and the policies that hampered growth, in fact, had precisely the opposite impact. The popular politics during the 1960s and 1970s showed bias towards "rural" areas where the voters lived. The industrial bias was towards "public" sector. Overall, the GDP grew at a snail's pace, and there was no real rise in per capita income.

During the 1980s, neighbouring China commenced its journey on market-oriented growth. At home, the governments of Indira Gandhi and Rajiv Gandhi began to tilt economic policy in the direction of big businesses. Foreign direct investment was still not a priority, but a few deals were successfully negotiated. The mantra of "garibi hatao" was retired to the Monopolies Restrictive Trade Practices Act, which made it hard for big businesses to invest in core sectors like chemicals, cement (Corbridge, 2009). The credit to large companies was made easier than ever before. Labour activism, which had become the face of industrialisation in the 1970s, was also tamed. Private sector investment was encouraged with some tax concessions. The private sector, though still small, began to contribute to economic development, while the public sector's role in capital formation had stabilised after a period of rapid growth in the 1970s.

The bigger push to embark on liberalisation in the 1990s did not come from the objective to reduce poverty; in fact, Deaton and Dreze (2002) point out that the rate of decline in poverty during the 1990s was not different from that during the 1980s despite government claims, but from the realisation that huge subsidies in and out of the agricultural system (fertiliser, electricity, water into and cheap food out of the public distribution system) would push the country deep into the fiscal and balance of payment crises that had erupted in 1991. To finance subsidies, the government had resorted to deficit financing and borrowing domestically and abroad, as the tax revenue was not sufficient. Tax concessions to big businesses in the 1980s combined with pervasive tax evasion did not help either.

The push for reforms in the 1990s also came from business communities who were tired of the pro-farming agendas of the National Front government led by Prime Minister V. P. Singh. Global economic thinking was also changing. USSR had collapsed. Margaret Thatcher had succeeded in dismantling public sector unions in the UK and privatising production in sectors that were largely public, and Ronald Reagan had succeeded in reversing many ills that the US was facing; he had inherited namely low growth, high unemployment and high inflation. The balance of payment crisis in 1991 provided the opportunity. Development economics was already out of fashion, and what was needed were sound monetary and fiscal policies and open trade and capital accounts.

The system of industrial licensing was dismantled in all but 18 industries and for locations other than 23 large cities with population above one million (Varshney, 1999). The foreign direct investment was encouraged in the Indian economy. This signalled India's new connection to the landscapes of globalisation. The telecommunication sector reveals more than any other industry the implications of liberalisation as the telecom revolution swept through middle-class India.

The relation of centre and state has also changed significantly since 1990. Prior to this, India's states had largely dependent on the centre, given the inelasticity of major state revenues, to seek extra funding as grants-in-aid under Article 275. Post-1990, the bargaining position of states has changed, and they are actively competing against each other to attract foreign direct investment or the funds of non-resident Indians. It is the states where the momentum of economic reform has taken root, which have used land acquisition and harsher labour laws as two major instruments to attract capital to their states. The centre is de facto encouraging states to free up extensive parcels of lands for deployment of private capital, and nearly 300 special economic zones (SEZs) that have come up since 2005 epitomise this attitude, which is a clear shift in economic policy from agriculture to non-agriculture. The Planning Commission reinvented its role in stark contrast to the past, from 'centralised planner' to 'indicative planner and policy maker'. The current government under Prime Minister Modi dissolved the Planning Commission and instituted a new entity called the National Institution for Transforming India (NITI) Ayog in 2015 to contribute to policy formulation.

In 1992, the model of decentralised governance gained further ground with the enactment of the 74th Constitutional Amendment Act. The Act devolves a number of urban functions to municipal bodies or urban local bodies (ULBs). However, the spirit of empowering local governments to determine their future was far from realised through the passing of the amendment (Tiwari, et al., 2015). Tiwari et al. (2015) argue that the "political centrality of cities" is not yet evident in India. The functional domain of ULBs have been steadily undermined by the state governments by setting up parastatals and diverting municipal functions (such as planning, and even provisioning of services such as water) and funds to them. State government, parastatals and UDAs continued to take over the functions that rightly belong to the ULBs (ibid).

In words of (Corbridge, 2009),

> What is now evident in India, even more so than previously, is a yawning gulf between the country's haves and have-nots. For the former, India is shining brightly. It is a land of Tata Nanos and shopping malls. It is a country that seems to be leap-frogging the industrial revolution to land talented people directly to those jobs – in IT, information processing and finance – that connect India to the globalizing world outside. This is precisely the land of SEZs, the Golden Quadrilateral, Gurgaon, the Bandra-Kurla complex in Mumbai, and various technopoles in Bangalore, Chennai and Hyderabad.
>
> (p. 19)

A major impact of advances in capitalism as it is occupying space has been that it is sweeping away those institutions that are likely to slow down the circulation time of capital. This is blatantly evident in unprecedented development activity that is taking place on urban fringes or peri-urban locations, and surrounds of Delhi are witnessing a major transformation.

Demographic turbulence

From a small city of 0.4 million population in 1901, Delhi has grown to become a mega city with 16.78 million inhabitants. Delhi witnessed a huge exodus of original landowners following massive land acquisition for the new imperial capital during 1911–20. However, these were replaced by in-migrants of the new capital. The announcement of Delhi as the capital of British Empire in 1911 added 18% population during 1911–21 and 30% during 1921–31. Other events, such as the war, brought in huge Allied troops to Delhi. The discontinuation of Shimla as summer capital of government in 1942 meant the capital was permanently occupied by government officials (Mehra, 2013). The announcement of the partitioning of the country in 1947 led to a huge displacement of people. About 300,000 Muslim residents of Delhi left the city, and half a million Hindu and Sikh migrants from the newly formed Pakistan came to settle in Delhi (Mehra, 2013). During 1941–51, the population of Delhi rose by 90%. In the decades that followed, labour migration from rural hinterland to urban has led to continuous growth of Delhi, and, on average, Delhi has added 50% to its population each decade (see Figure 2.1).

The impact of population growth on its spatial distribution has been phenomenal. From a walled city of Shahjahanabad, the city has grown manifolds. In a major interactive online project aimed to explore stories from 1947 by the *Hindustan Times*, Dawn and the 1947 Partition Archive, Alluri and Bhatia (in press) present a cartographic representation of the change in occupied space in Delhi between 1942 (pre-independence) and 1956 (a decade after independence). The period of 1947–56 was a period of major resettlement of migrants from across the border. The resettlement colonies that were developed replaced the

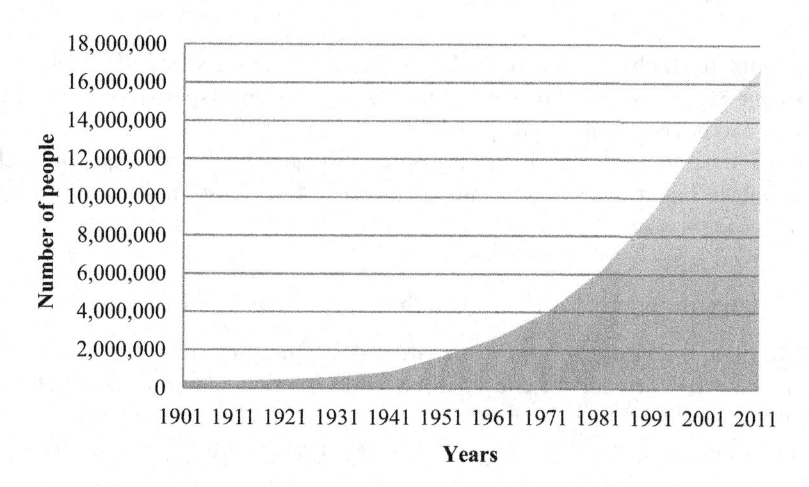

Figure 2.1 Population of Delhi (1901–2011)

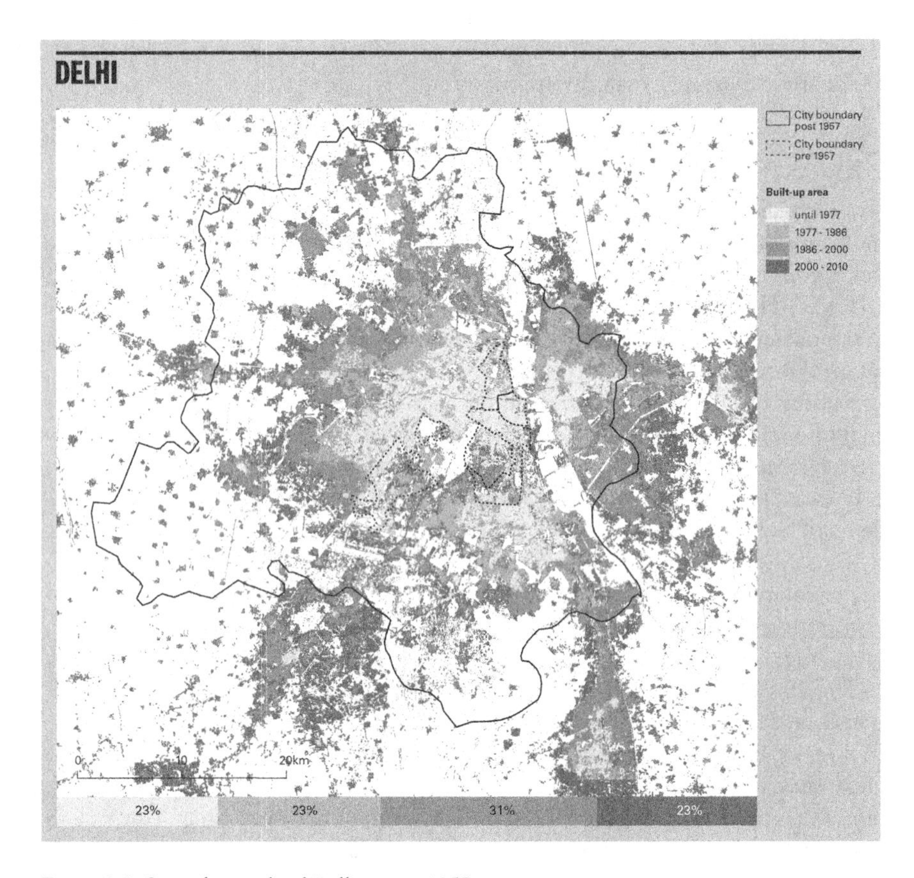

Figure 2.2 Spatial growth of Delhi since 1957

Source: LSE Cities (2014)

hutments in which migrants had taken shelter. The Lutyens Delhi (New Delhi) to a large extent was not touched, but resettlement colonies sprang all around the city, as can be seen from the map of 1956 (see Figure 3.9).

From 1957 onwards, the built area of the city grew by almost a quarter each decade (see Figure 2.2) to accommodate the 50% increase in population that each decade ensued.

2.6 Managing Delhi's urban growth prior to master planning

For India and Delhi, the transition into a democratic republic nation was far from smooth. Partition of the country, violence that it ensued and refugee crisis that followed meant that many regulations and institutions of central control and associated bureaucracy of British legacy continued well after the independence. The refugee crisis that began on the day of independence meant that housing shortage and lack of infrastructure reached catastrophic levels. Makeshift camps with temporary structures on any open land owned by government (including Purana Qila) were constructed. There were encroachments on public spaces as the migrants opened up temporary shops to sell goods to make their both ends meet, defying any orderly planning that New Delhi had so jealously guarded. Migrants complained of high rents that the government had been demanding for the shops that were constructed for them. They also complained of widespread corruption in government machinery that was involved in management of refugee camps (Mehra, 2013). The new democratic government argued for centralisation of policies and resources to address the crisis in hand, in the same way as the British rulers had justified as a vision for British India. Laws such as Delhi Refugees Registration Ordinance 1947, to register refugees and provide resettlement benefits only to these registered, were enacted. The intention was to control the numbers of migrants to Delhi after Partition. Deteriorating living conditions and concerns about health led to demands for building new townships to house migrants (Mehra, 2013).

Efforts to resettle migrants led to the development of new townships by the Ministry of Rehabilitation. Government-owned land, land that was in custody of DIT and freshly acquired land were utilised for this purpose. The Ministry built new houses and developed sites as town extension schemes and secondary hinterland towns. These developments adhered to planning norms. The mechanism for the allocation of plots and houses was through direct sale, or sale through cooperative societies, or through tendering/auctioning of the leases (Mehra, 2013). These plots were sold on 99-year leases and were to be developed only for permitted 'land use'. The government of India became one of the largest real estate owners and developers in the city (ibid).

During the 1950s, the governments of India and Delhi framed legislation, the Government Premises (Eviction) Act, to clear public lands from encroachments that had appeared after Partition as migrants undertook unauthorised construction. Protests that ensued led government to promise regularisation of these encroachments. Despite these promises, programmes to clear slums continued for many decades.

By the mid-1950s, Nehru's capitalism had gained ground, and with that came the migration of professionals into white-collar jobs and bureaucracy. The construction activity that the development caused led to the migration of construction workers from surrounding areas. Overall housing shortage had compounded manifold. Unable to find formal shelter, low-income workers squatted on public lands; those who were slightly better-off undertook unauthorised construction, and unregulated private colonies on private land came up. Some Delhi entrepreneurs started constructing housing colonies on their own lands or agricultural land on urban fringes as a response to the opportunity that housing shortage had presented. The period 1955–59 was "a period of boom for private land development companies and house construction" (Bose, 1973).

2.7 The "Master Plan" era

In 1953, the States Reorganisation Commission decided that Delhi would remain under the control of the Union government as a union territory (UT) considering its special status as a capital (Sasidharan, 2015). In the same year a parliamentary bill to establish the Municipal Corporation of Delhi (MCD) under the Delhi Municipal Corporation Act of 1953 was introduced (ibid). A number of new government institutions to 'regulate' rampantly growing disorderly urban built space emerged. The approach was top-down central government led. The Ministry of Health inherited the DIT in its portfolio. The notion of Greater Delhi encompassing areas of New Delhi, 'Old Delhi' and newly developed fringe townships came about. The Town and Country Planning Organisation, established in 1955, prepared a comprehensive interim plan for Greater Delhi in 1956. A master plan for Delhi's future growth was prepared in the same year, which envisioned Greater Delhi as a greatly expanded, rationalised metropolis set within a regional framework, with New Delhi at its centre (Mehra, 2013).

Parallel to the preparation of the Delhi master plan, attempts to control haphazard development activities through a series of provisional measures to regulate new private buildings were made. In 1955, the Delhi Control of Building Operations Ordinance was instituted "to prevent unauthorised construction in the open land in and around the urban area of Delhi". Some areas, including the land within and outside of DMC and New Delhi municipality were declared as 'controlled' (Mehra, 2013). A new authority under the Act, the Delhi Development (Provisional) Authority (DDPA), was established with the mandate to oversee the growth of these 'controlled areas'. The authority was also required to ensure these new developments complied with building by-laws and proper planning.

All new buildings in 'controlled' areas required permission from the DDPA. Consequence of such control was that the development shifted to the fringes in border states such as Uttar Pradesh. Many private developers started developing buildings in areas between Gazhiabad in Uttar Pradesh and Shahadara, as there were no regulatory controls hindering their activities there. The impact of the heavy-handed regulatory regime was that it started stifling the activity of private-sector developers who had emerged as a response to slow public machinery in meeting housing shortages, and these regulations, instead of curbing urban sprawl, caused them, albeit on urban fringes in border-states where Delhi's regulatory regimes did not apply. The overall impact was speculation, land hoarding by private developers of agricultural land in rural hinterlands and haphazard and poor development often in connivance with government officials (Mehra, 2013). The number of 'unauthorised' squatters in Delhi was estimated to be around 200,000 (ibid).

In 1957, the Delhi Development Authority (DDA) was established through an act of Parliament to replace the DDPA and the DIT. DDA was tasked with promoting and securing the development of Delhi in accordance with the Master Plan through an explicitly top-down process (Sheik & Mandelkern, 2014). Besides enforcing the city's zoning and building codes as defined by the Master Plan, the

DDA monopolised all future land ownership and development in Delhi, thereby extending government control over tens of thousands of acres in Delhi's rural hinterland. The Master Plan for Delhi-1962 (MPD-1962), which covered the period 1962 to 1981, was based on a regional approach with recommendations for the Delhi Metropolitan Area (DMA) (Sasidharan, 2015). Since the context for the plan was to curb speculation of land and restrict private land development, the acquisition of land by the government ironically implied that the control of land could be a good instrument for managing city growth (Sasidharan, 2015).

The implementation of the Master Plan led to the displacement of the economic centre from Old Delhi, composed of small markets and businesses, to South Delhi (Ahmad et al., 2013). The planning approach for the plan was to limit the growth of core city by establishing green belts around the city. A decentralised model of spatial development with the development of seven neighbouring towns was proposed (Ahmad et al., 2013). (Ahmad et al., 2013) argues that this initiated the low-density urban development model in Delhi and caused urban sprawl. Slums were deplored, and the approach towards slums was to clear them and resettle their inhabitants on urban fringes where price of land was low. The socialisation of Delhi's land market commenced with 30,800 hectares of land notified under the Land Acquisition Act (Sasidharan, 2015). The land policy was to acquire land by force, assemble it for future urban development and dispose the same on a leasehold basis (ibid). Though the policy had novel intentions for coordinated growth, through an apex planning, land development and controlling authority, it caused numerous unintended outcomes: unprecedented growth in land prices; artificial scarcity of land for housing; unauthorised land development; and the proliferation of informal settlements in the city (Sasidharan, 2015). The irony was that the DDA became the planning and development authority under one umbrella. A discerning critic would argue the extent of regulatory capture that such a situation would guarantee and the fear is not baseless. Sheik and Mandelkern (2014) report that by 1968, the staff strength of DDA was as high as 35,000 spread over a construction division, a building division and a land acquisition division and included planners, engineers, horticulturalists and other staff.

The second Master Plan of Delhi (1981–2001) continued the earlier philosophy of public-sector-led growth and development. Most of Delhi's urban areas were developed by DDA using this method for four decades, and during this phase, the "entire financial benefits of converting peri-urban agricultural lands to high value land for urban development was captured by DDA, without delivering on the original agenda of such a social endeavour" (Sasidharan, 2015).

Bose (1973) argues that the DDA, like its predecessor DIT, ignored the housing programme and concentrated on the development of land with the prime objective of maximising profits on this disposal of land. Not much housing was developed, causing persistent housing shortages, particularly of affordable housing. The outcome of such planning was nothing but unintended urbanism, despite the government's monopoly on land and building control (Bose, 1973).

The third Master Plan 2001–21 (though it was approved in 2007 by the Ministry of Urban Development after a series of public consultations) proposes the involvement of the private sector in assembly and development of land along with provisions for adding 22,000 hectares of urbanisable land. However, in the recent years, opposition to compulsory land acquisition has mounted, and this is leading DDA to explore other models for land assembly, in particular models related to land pooling.

Strong involvement of government agencies in the land market since independence has created distortions to an extent that the large private developers have largely exited from development activities in Delhi, and they have been working outside Delhi's boundaries in the neighbouring cities of Gurugram (Gurgaon), Noida and Faridabad. The spatial patterns of jobs and residential locations that have emerged during 1980–2000 period have been dumbbell-shaped, where a large population is located in fringe locations, owing to the availability of housing in these locations, while jobs were concentrated in Delhi.

During the past six decades, a quagmire of institutions that govern various aspects of Delhi have emerged, which has caused confusion in planning and development of the city (Ahmad, 2013).

2.8 National Capital Region – Delhi in a regional context

The phenomenal population growth in Delhi after independence and prevailing planning ideological paradigm of 1960s (as reflected in the 1956 Interim General Plan and reiterated in the Master Plan of Delhi 1962) to decongest Delhi led to the enactment of National Capital Region Planning Board Act in 1985 to plan Delhi in the regional context. A new agency called the National Capital Regional Planning Board (NCRPB) was set up with the mandate to prepare a regional plan for National Capital Region, now comprising 22 districts from four surrounding states besides the National Capital Territory of Delhi. Much of the growth due to the spillover of the population initially (and jobs later) has been in areas abutting Delhi – Gurugram, Faridabad, Noida and Ghaziabad. While a detailed discussion of the Regional Plans of 2001 and 2021 prepared by NCRPB is beyond the scope of this chapter and the book, it may suffice to highlight that these regional plans aim to promote growth and balanced regional development strategies (NCPRB, 2005). The Regional Plan, as required by the NCPRB Act 1985, formulates proposals related to land use and land allocation for various uses, urban settlement patterns, transport and communication for the NCR, water and sewerage, and identifies areas that require immediate development attention. In doing so, it divides the region into National Capital Territory of Delhi (NCTD), the Central National Capital Region (CNCR) comprising immediate peripheral districts (Gurugram, Faridabad, Bahadurgarh, Noida, Ghaziabad and Sonepat) of NCTD and the rest of the NCR.

The Regional Plan 2021 has proposed development of regional transport linkages, modal industrial estates and SEZs in the CNCR and wider NCR regions.

The Regional Plans have faced numerous challenges in their effective implementation. The planning has lagged behind urban trends. Despite the proposals in the Regional Plan 2001 to decongest NCTD region by developing CNCR, the population of NCTD continues to grow at an unprecedented rate. The Master Plan of Delhi 2021 saw withdrawal of DDA from development activities and a larger involvement of private developers, as envisaged in the Master Plan, in redevelopment of existing stock to higher permitted density. The pace was, however, slower due to the limited supply of land (and inactivity of DDA in land procurement) than required to meet the demand. The repeal of the Urban Land (Ceiling and Regulation) Act 1976, which had restricted landholding by landowners in urban areas giving the power to the Act to acquire land deemed surplus, further boosted the construction activity in CNCR as states such as Haryana were first to repeal the Act (Sasidharan, 2015). A number of SEZs and industrial estates have come up largely in the CNCR region to take advantage of its proximity to Delhi. The land for these has largely been acquired using the Land Acquisition Act 1894. The huge difference between the compensation paid for market price of the land and the forcible nature of acquisition (often way higher than the requirements) have led to protests, which, in many instances, have been violent (Kennedy, 2014).

There have also been challenges in funding proposed infrastructure development in NCR, as it required financing to come from all three levels of government for various projects having a sub-regional, regional or national interest. The Regional Plan also proposes private-sector investments through public-private partnerships or otherwise, but this has been very limited.

2.9 Current state of Delhi

The economic reforms that began in 1991 in the country led to continued migration of people in Delhi and its surrounds. In the same year, Delhi was declared as quasi-state under the name National Capital Territory of Delhi (Sasidharan, 2015). The new status gave a legislative assembly and elected government for Delhi with powers to make legislation on all matters except law and order and land. Land still remains a subject matter of central government that controls it through the lieutenant governor, who is also chairman of DDA (ibid). The land policy requires cooperation between state and central governments, which has become a contentious issue given ideological differences of political parties governing at central and state levels.

Delhi of the twenty-first century is a complex urban space where congested 'old' co-exists with 'controlled' low-density 'new' Delhi side by side. There are pockets, which are recognised as 'villages' but are engulfed all around with highly urbanised spatial developments. There are many such dualities that Delhi's spatial structure presents – urban-rural, public-private, legal-illegal, built-open, protected-free – which are often pitted against each other by administrative institutions of the state to make, in the words of Sheth (2017), a "patchwork quilt at best". This section briefly touches upon these dualities to emphasise the current state of Delhi and its surrounds.

2.9.1 Housing shortage

In an assessment of the performance of Delhi's development and planning agency, Sheik and Mandelkern (2014) appropriately argue that DDA has been inefficient in building and allocating low-income housing despite being so successful in acquiring land and providing high-end amenities. The overall performance of the DDA, as highlighted by Sheik and Mandelkern (2014), has been abysmal with only a quarter of population in Delhi living in planned colonies by the turn of millennium. The bureaucratic administrative set-up of DDA has created a distance between those who plan, those who execute and those who witness the implication of planning on ground (ibid). As (Sheik & Mandelkern, 2014) aptly put it, "the result is an agency that plans, builds, and maintains an 'aesthetic' city, one that privileges parks over functional infrastructure, cleanliness over liveability and a 'world class' veneer over inclusion".

During the first two decades of its operation (the period of the first Master Plan of Delhi), DDA acquired large tracts of land but failed miserably on its commitment to deliver developed residential units for which land was acquired. The approach to implementation of the vision of planned city was negative, and the agency resorted to removing unauthorised colonies and *jhuggi jhompri* (temporary hutments) clusters from public lands, which housed 30% of the population, rather than building low-income housing. Political upheavals such as imposition of emergency in 1977 caused delays in preparation of the second Master Plan of Delhi. When finally, four years after the expiration of first plan, it got prepared, it did not get approval, as it was seen to be in conflict with the regional plan that the National Capital Region Planning Board, established in 1985, had prepared. The overall implication was that the housing shortage continued to manifest itself, and DDA could meet only one-third of the demand.

The 1990s ushered in an era of economic reforms in India, and this implied that the approach to land and housing also changed. Government policies moved towards reducing control on land and housing. DDA reformed its tenure policies away from leasehold and towards freehold (Sheik & Mandelkern, 2014). While this resulted in more mature system of tenure which allowed owners of residential units to leverage asset value of their houses, it also put pressure on real estate markets, further marginalising the poor. DDA sought to auction its land bank for commercial development, and the encroachments stood in the way of realizable financial gains (ibid).

The Third Master Plan (2001–present), recognising the shortcomings of earlier plans and their inability in addressing housing shortage, proposed a paradigm shift where the government sought to deliver a better-planned city through public-private partnerships. In their assessment of DDA's performance since 2001, Sheik and Mandelkern (2014) highlight that DDA's role and landowning agency has continued unabated, while its contribution to the construction of housing units for the weakest segment of society has been marginal. During 2003–10, only half of the targeted housing units had been completed, and of the total new housing completed, DDA's contribution had been less than 20%.

2.9.2 Rural-urban dichotomy

The first official land survey of Delhi in 1908 identifies the land outside city limits as villages and records them as agricultural (*khasra*) land and inhabited (*abadi*) land. While each parcel of the agricultural land had been recorded separately and had a separate number, the whole of inhabited land in a village had been recorded on a single *khasra* number. On a survey map, inhabited rural areas were encircled with red lines; hence these came to be called as Lal Dora areas. Properties in these areas were transferred in hereditary lines rather than through formal transactions. A formal record of ownership did not exist, and with time this had become unclear, as properties had been subdivided, transferred, sold without any formal registration. When the mega capital city development programme began in 1912, rural lands on urban fringes were acquired to build viceregal residences and British government of India offices. Later, the land was acquired to build resettlement colonies for migrants who took refuge in the city after Partition. A total of 47 villages were acquired and urbanised between 1912 and 1951 (Sheth, 2017). The acquisitions included *khasra* and *abadi* areas and led to displacement of more than 5,000 inhabitants. During the 1950s, planners needed more land to construct houses to meet the shortage, which had reached a colossal scale, but the procedure for acquiring land from villagers was completely reformulated (ibid). The democratic government could not displace the inhabitants of villages. Villagers were coerced in selling their farmlands willy-nilly at government-authorised rates, but they resisted their residential quarters from being taken (Sheth, 2017). The government acquired the land around habitation and left the inhabited area untouched. The process of land acquisition by DDA took 10 to 15 years from the date of serving the land acquisition notice to physical possession, and during this time the inhabited areas were officially designated as "urban villages". As part of compensation, DDA gave a 400-square-metre residential plot in other parts of the city to the acquiree; by the early 1980s, the size of the alternative residential plot was reduced to 250 square metres.

The urban villages were treated differently, as they did not become part of municipal corporation, and the building by-laws of DDA did not apply. The authorities used the *khasra* and *abadi* maps from the 1908 land survey of Delhi as their blueprint for implementing these exceptions. The 1908 land survey became the pseudo-legal historical precedent by which the exclusivity and rights of urban villagers were justified, and perhaps even valorised, in city planning discourse.

2.10 Noida, Gurgaon, Faridabad, Ghaziabad – the modern surrounds of Delhi

Two major impacts of the urban pattern in Delhi and its surrounds can be observed in the Central National Capital Region of Noida, Gurgaon, Faridabad and Ghaziabad.

First is the distribution of urban population growth across different regions in NCR. As presented in Bedi (2014), the population of the National Capital

Territory of Delhi grew almost three times, from 5.7 million to 16.3 million, during 1981–2011. During the same period, the Gurugram-Manesar area of CNCR grew from 0.1 to 0.9 million and Faridabad-Ballabhgarh grew from 0.36 to 1.4 million in population. The towns of Noida and Ghazianad-Loni area witnessed similar high growth. The population of Noida grew from 36,000 in 1981 to 0.6 million in 2011. Ghaziabad-Loni grew from 0.3 million in 1981 to 2.1 million in 2011. While Delhi had the advantage of planned growth, these peripheral areas (except Noida) were largely growing in an unplanned way, driven by market forces led by private developers.

Second is the growth in built-up area, which has witnessed phenomenal growth resulting from private sector-led construction activity on the peripheries of Delhi, in other states largely due to strong control on land supply in Delhi. Private developers have acquired large agricultural lands in Gurugram and Faridabad for development purposes (Bedi, 2014). In addition, states on the fringes of Delhi have acquired and designated lands for development of SEZs (Kennedy, 2014), causing further alienation of land from villagers.

2.11 Conclusion

A summative assessment of the market, policy and regulatory review of Delhi over 2,000 years presents an important context within which the urban form and structures could be understood. The chapter presents the 'market', 'agency' and 'processes' within which the built form that we see today evolved. What is interesting to note is that in Delhi, the control over land has remained in the hands of ruler (prior to independence)/government (after independence) irrespective of whether they were monarchs or are elected representatives. What has been built is a reflection of their ideas, ambitions, arrogance and appreciation. What comes out of the review is that while the city witnessed many changes in its rulers, the continuity in ideas was provided by the bureaucracy that existed in all periods irrespective of whether they were working for Sultanate rulers, Mughal rulers, British or the modern Indian democratic government.

Earlier Sultanate rulers adopted economic systems of their Hindu predecessor albeit with minor modifications but provided continuity. The Mughals and British retained the system of intermediaries in the collection of land revenue of the Sultanate period or their predecessors. The land tax system that Khilji designed during twelfth century has influenced the land tax system even in modern India. The rulers of Delhi were obsessed with building new cities, as they amassed wealth. There is a high degree of correlation between the economic condition, political stability and the constructed built space, as discussed in Chapters 3 and 4. Shahjahanabad of the seventeenth century was the most elaborately planned city, dotted with buildings and spaces that reflected the economic and political stability of the Mughal Empire.

For a brief period, Delhi lost its status as the capital when the British captured Delhi in 1857 and cautiously transformed Shahjahanabad into a commercial centre. The land tenure underwent changes with permanent settlement with zamindars, but it never was a "freehold". When, in 1911, New Delhi was declared as the capital of British India, new construction began on a land that was compulsorily acquired.

The migration that followed the partition in 1947 and economic development led to an expansion of city largely built through instruments that socialised private land through compulsory acquisitions. Much of the development in Delhi either during pre-independence or post-independence periods has ignored the poor. Even during democratic rule after 1947, Delhi pursued the policy of eviction of those who lived in informal settlements despite that the boundary between formal and informal was often illusive. The formal instruments of planning made the living conditions far more precarious as they were delinked with the ground reality. Multiplicity of governments and its institutions and stubborn resistance in involving the private sector in any matter related to land development for long worsened the situation with the consequence of resulting housing shortage, lack of land supply and further marginalisation of the poor.

The "surrounds" is an important connotation. Whenever Delhi felt the need to address its problems in the existing city, it built new by encroaching its surrounds. All cities of Delhi have reflected this attitude throughout its history. Even post-independence Delhi has expanded into its surrounds to meet its requirements despite the fact that the areas in New Delhi are low density compared to its peer cities of a similar population.

Notes

1 Where the land revenue was imposed indirectly – through agreements made with zamindars (big landlords) – the system of assessment was known as zamindari. Under the permanent settlement system between the British East India Company and the zamindars, (1) zamindars were recognised as proprietors of the soil with rights of free hereditary succession, sale and mortgage. However, these rights were conditioned to the timely payment of land revenue failing which the land was acquired and ownership rights were ceased; (2) the system stipulated that the zamindar should safeguard the rights of their tenants by granting those *pattas* (lease document) or documents stating the area and rent of their respective holdings; and (3) the zamindars were made subject to such rules as might be enacted by the government for securing the rights and privileges of the tenants in their respective tenures and for protecting them against undue or oppression. All *abwabs*, or cesses levied by the zamindars in addition to the rent, were abolished. The transit duties and road and ferry tolls were taken over by the government, but the market tools and profits from fisheries, trees and waste land were left entirely to the zamindars (Husain & Sarwar, 2012).

2 This system was instituted in some parts of British India by 1820. Under the *Raiyatwari* system, tax was assessed annually ('annual settlement'), and every registered holder of land was recognised as its proprietor and was paying rent directly to the government. He was at liberty to sublet his property or to transfer it by gift, sale or mortgage. He could not be ejected by the government so long as he paid the fixed assessment and had the option annually of increasing or diminishing his holding or of entirely abandoning it (Husain & Sarwar, 2012).

3 The *Mahalwari* system was introduced by 1822 with the estate or "mahals" proprietary bodies where lands belonged jointly to the village community technically called the body of co-shares. The body of co-shares was jointly responsible for the payment of land

revenue though individual responsibility was not left out completely and head of village called *lambardar* was consulted in fixing the land revenue rate.

4 The North-Western Provinces were created in 1836 when ceded and conquered provinces (1801 and 1803, respectively) in the north got merged with the western provinces. Oudh came under the administration of the North-Western Provinces in 1843–44, and from that time it has been named as the North-Western Provinces and Oudh. The name was changed to United Provinces in 1901–2. It included the whole portion of present Uttar Pradesh with some portions of Madhya Pradesh and Haryana and also includes the whole area of Uttarakhand and Delhi (http://shodhganga.inflibnet.ac.in/bitstream/10603/13529/6/06_chapter%201.pdf).
5 The Company official in charge of land-revenues at the district-level also had judicial authority in the event of disputes.
6 *Khasra* is the unit number assigned to a specific plot of land.

Bibliography

Ahmad, S., Balaban, O., Doll, C.N.H. & Dryfus, M. (2013). Delhi Revisited. *Cities*, *31*, 641–653.

Alluri, A., & Bhatia, G. (Ongoing). The Decade That Changed Delhi. *Hindustan Times and Dawn*. New Delhi.

Bedi, J. (2014). *Urbanisation, Development and Housing Requirements in the National Capital Region (NCR)*. New Delhi: National Council of Applied Economic Research (NCAER).

Bhagwati, J. (1993). *India in Transition: Feeling the Economy*. Claredon: Oxford University Press.

Bose, A. (1973). Land Prices and Land Speculation in Urban Delhi 1947–67. In A. Bose (Ed.), *Studies in India's Urbanisation*. New Delhi: IEG.

Corbridge, S. (2009). The Political Economy of Development in India Since Independence. In P. Brass (Ed.), *Handbook of South Asian Politics*. London: Routledge.

Deaton, A., & Dreze, J. (2002). Poverty and Inequality in India: A Re-Examination. *Economic and Political Weekly*, *37*(36), 3729–3748.

Guha, R. (1996). *A Rule of Property for Bengal: An Essay on the Idea of Permanent Settlement*. Durham, NC: Duke University Press.

Husain, D. M., & Sarwar, F. H. (2012). A Comparative Study of Zamindari, Raiyatwari and Mahalwari Land Revenue Settlements: The Colonial Mechanisms of Surplus Extraction in 19th Century British India. *Journal of Humanities and Social Science*, *2*(4), 16–26.

Johnson, D. (2015). Land Acquisition, Landlessness and the Building of New Delhi. In D. Johnson (Ed.), *New Delhi: The Last Emperial City* (pp. 161–195). London: Palgrave Macmillan.

Kaviraj, S. (1984). On the Crisis of Political Institutions in India. *Contributions to Indian Sociology*, *18*, 223–243.

Kennedy, L. (2014). Haryana: Beyond the Rural Urban Divide. In R. K. Jenkins (Ed.), *Power, Policy and Protest: The Politics of India's Special Economic Zones*. Oxford: Oxford University Press.

Lewis, W. A. (1954). Economic Development with Unlimited Supplies of Labour. *The Manchester School*, *22*(2), 131–191.

LSE Cities. (2014). *Patterns of Growth, Delhi*. Retrieved from Urban Age/LSE Cities: https://LSECiti.es/u164112b6

Mehra, D. (2013). Planning Delhi ca. 1936–1959. *Journal of South Asian Studies*, *36*(3), 354–374.

MoUD. (2007). *Report of the Expert Committee on Lal Dora*. New Delhi: Ministry of Urban Development.

NCPRB. (2005). *Regional Plan 2012. National Capital Region Planning Board (NCPRB)*.

Panagariya, A. (2008). *India: The Emerging Giant. New York: Oxford University Press.*

Pati, S. (2015). *The Regime of Registers: Land Ownership and State Planning in Urban Villages of Delhi*. London: University of London, South Asia Institute.

Pearson, M. (1985). Land, Noble and Ruler in Mughal India. *Sydney Studies in Society and Culture*, 175–196.

Raychaudhuri, T., Habib, I. & Kumar, D. (1982). *The Cambridge Economic History of India, Vol. 2*. Cambridge: Cambridge University Press.

Sasidharan, S. (2015). *Landlocked in Peri-Urban Politics Around Delhi's Land Policy*. Paper presented at RC21 Conference, 19–27 August 2015, Italy.

Sheik, S. & Mandelkern, B. (2014). *Delhi Development Authority: Accumulation Without Development*. New Delhi: Centre for Policy Research.

Sheth, S. (2017). Historical Transformations in Boundary and Land Use in New Delhi's Urban Villages. *Economic and Political Weekly*, *LII*(5), 41–49.

Swamy, A. V., & College, W. (2010). Land and Law in Colonial India. In D. Ma, & J. L. Zanden (Eds.), Long-term Economic Change in Eurasian Perspective (p. na). Stanford, CA: Stanford University Press.

Tiwari, P., Nair, R., Ankinnapalli, P., Rao, J., Hingorani, P. & Gulati, M. (2015). *India's Reluctant Urbanization: Thinking Beyond*. London: Palgrave Macmillan.

Varshney, A. (1999). Mass Politics or Elite Politics? India's Economic Reforms in Comparative Perspective. In J. Sachs et al. (Eds.), *India in the Era of Economic Reforms*. Oxford: Oxford University Press.

3 Evolution of urban form and imageability

The culmination of a new 'city' begins with the start of a new pattern of settlement, society, culture, religion, language, art and architecture. The new elements are often woven over the existing fabric, thus giving it the newness while also preserving the original. The city of Delhi contains many layers of human settlements that existed at different points in time and are woven together into its built environment as it exists today. The evolution of a city is in itself a complex process, and the existence of the old strata underneath that makes it much more complex and thus also interesting to read.

There is a growing curiosity to understand how Indian cities like Delhi, with superimposed layers of histories of physical and traditional elements, come to terms with modern theories of urban planning and design. Many urban theories are constructed to explain the form and formation of urban centres. As an outcome of industrialisation, urbanisation was a popular phenomenon of the West, and most of these theories are therefore contextual to Western urban centres. While other parts of the world are developing, there is a growing curiosity to study new emerging urban centres. However, the existing theoretical frameworks pose challenges to their direct application to non-Western cities, and there is a need to develop contextual frameworks. With this objective, this chapter starts by explaining existing theoretical frameworks. Towards the end of this chapter, these discussions are looked at again to include components of Indian cities.

Before entering into a discussion on urban form, it is considered necessary to derive an appropriate definition that can make possible the discussion on cities from ancient, medieval and modern times. The first two sections provide the theoretical framework for further discussions on urban form. While the contemporary definition[1] of the term 'urban' is simple to understand, it needs to be widened to create the scope for historical cities, which are an integral part of modern cities, like Delhi. Therefore, the chapter starts by discussing historians' approach to defining 'urban' and analysing its determinants. Also, there is a growing urge among design disciplines, such as architecture, urban design and urban planning, to understand the agents that determine urban form of cities. This furthers the urgency of deriving a combined framework within which multiple factors of the complex city-building process can be studied in parallel to its physical form and design.

The discussions in this chapter shall be limited to the understanding of urban form of the region, and the following chapter should cover the details of architectural elements that characterise different time periods in the history of Delhi.

3.1 Defining 'urban'

'The concept of "city" is notoriously hard to define' (Childe, 1950, p. 3). In his widely cited publication, 'The Urban Revolution', Gordon Childe (1950) derives an abstract criteria for defining 'city', or rather a prehistoric city. He identifies three stages of human evolution, the first stage being 'savagery', followed by 'barbarism' and 'civilisation'. While the first two stages are conveniently defined by the methods of procuring food from hunting-gathering and domestication, respectively, the third stage of 'civilisation' is complicated in definition. Childe (1950) writes that "etymologically the word (civilisation) is connected with 'city', and sure enough life in cities begins with this stage". Thus, Childe (1950) viewed 'city' as a stage of human evolution. Based on the information available from prehistoric settlements of Egypt, Indus, Sumeria, and Maya, the following ten criteria were derived:

i. Size and population density

Childe (1950) writes that a certain size and population density is essential prerequisite for a settlement to be considered a city. Among these prehistoric sites, the size of settlements in Mesopotamia and Indus Civilisation is calculated with some reasonable confidence. It is estimated that the population of Sumerian cities (in Mesopotamia) ranged between 7,000 and 20,000, and the cities of Mohenjo-daro and Harappa (in Indus) were probably on the higher side. As per Lambrick's calculation, the approximate population of Mohenjo-daro city is estimated to be 35,000 people (Allchin & Allchin, 1968). Allchin and Allchin (1968) expect the population of Harappa to be similar to that of Mohenjo-daro, given that the two cities are comparable in size.

Most probable reasons for bigger size of these cities, as supposed to villages, are often linked with the availability of natural resources, the techniques of exploiting these resources, transportation mechanism and the method for storing and preserving food (Childe, 1950).

Allchin and Allchin (1968) write that the sudden rise in the size of settlements and population in Indus Valley Civilisation, as compared to the earlier settlements on neighbouring sites, was an "outcome of successful control of the tremendous agriculturally productive potentials of the Indus plains" (p. 126). Being located on the flood plains of the Indus River, there was always a fear of flooding, and, therefore, ramparts were constructed all around the city to protect the houses and grainaries against floods (Allchin & Allchin, 1968).

ii. Agricultural surplus and emergence of specialised trades

The composition and function of cities differ from that of village in regard to the nature of employment of the citizens (Childe, 1950). Childe (1950) explains that production of agricultural surplus allowed exchange of food grains for specialised services from craftsmen, bricklayers (in Indus), transporter, merchants, officials and priests. This encouraged a few citizens to pursue full-time employment in specialised trades. These specialists did not grow their own food and were rather dependent on the surplus produced by peasants from the city and dependent villages. The diversification of economic activities gave rise to new trades, thus also changing the social composition and economic functions of the urban population.

iii. Payment of tax on the surplus

Given the technological limitation in prehistoric times, there was little surplus produced from land. However, the citizens were concentrating a portion from the surplus with the king or to the deity, in the form of tax or donation. Childe writes that "without this concentration, owing to the low productivity of the rural economy, no effective capital would have been available" (pp. 11–12).

iv. Monumental public buildings

The presence of monumental public buildings in cities makes them physically different from villages. The monuments are also symbolic of concentration of "social surplus" or food grains (Childe, 1950). To explain more, Childe (1950) gives example of temples in Sumerian cities which also housed an enormous granary. Another example is the citadels of the Harappa and Mohenjo-daro cities that are built on a raised podium built from mud-bricks, overlooking big granaries.

v. Ruling class

Childe (1950) suggests that the non-agriculturists were supported on the agricultural surplus accumulated by the temple or the court. Efficient social organisation was required for harmonious economic exchanges, thus demanding better administration than what a 'tribal chief' of Neolithic societies could provide.

vi. A system of recording and measuring

Economic activities like trade, tax and revenue collection compelled invention of units of measure and methods of recording, through writing, that is understandable by colleagues and successors. However, the characters used for writing varied across regions of Egypt, Mesopotamia, the Indus Valley and Central America.

vii. The use of script

Childe (1950) writes that the use of script gave way to predictive sciences – arithmetic, geometry and astronomy. In Egyptian and Mayan civilisation, the script could be used to create calendar that was useful for the rulers to regulate the production cycle of agriculture (ibid).

viii. Sophisticated style of carving and drawing

Full-time employment of sculptors, painters and other craftsmen led to the invention of sophisticated styles of art and craft in different regions. The style of painting and sculpture was much more advanced than from the abstract drawings of the Neolithic ancestors.

ix. Foreign trade

Foreign trade over long distance was common in early civilisations. Trade goods included not only luxury items but also essential goods and raw materials for industrial production. Also, a portion of social surplus was paid for the importation of raw materials.

x. Security from the state

The specialist craftsmen were provided with raw materials necessary for their employment as well as were guaranteed a security in the state organisation. This system was recognising the residential association of these specialists with the state and was less considerate of their kinship.

The preceding criteria majorly look into economic, political and social organisation of the society as the prerequisite for its settlement to be called 'urban'. Put another way, these points differentiate a tribal settlement from a 'civilised' settlement. And in doing that, a 'civilisation' is assumed to be residing in 'urban'.

The Indian history of civilisation starts with Indus Valley Civilisation of 2500 BC.[2] This settlement exceeds Childe's (1950) expectations in many ways. First, the built form is spectacular and, there was a remarkable uniformity in physical layout of cities, towns and bigger settlements of the Indus Valley Civilisation (Allchin & Allchin, 1968). Common elements of all planned cities of Indus include the following (Allchin & Allchin, 1968):

- Regular orientation;
- High citadel on the western side of the city, overlooking the lower city;
- The city was mostly squarish;
- Roads were laid out in a grid pattern, intersecting at right angles and circumscribing the blocks of dwellings. The size of roads was carefully decided by a modulus;
- There was standardisation of sizes of bricks, both burnt bricks and mud bricks;
- The precision of brickwork at the Great Bath at Mohenjo-daro and granaries in both the cities (Mohenjo-daro and Harappa) demonstrate highly skilled bricklayers;
- The brickwork of houses is regular and monotonous, without any decoration;
- There are bathrooms and latrines in the houses, which are connected, through chutes and covered brick drains, with soakage pits and sumps.

There is limited understanding on non-physical characteristics of Indus Valley, and with some reservations Allchin and Allchin (1968) explain the cultural, social and political structure of the Indus Valley Civilisation, as inferred from its physical remains. For example, the difference in the size of houses and the grouping of "baracks" (for labourers) at a particular location probably indicated the presence of social hierarchy and class differences. The same evidence may also indicate a "caste" system, as observed in later times. The presence of citadel, public granaries, and seals gives an indication of some form of administration, either priest-kings or at least a priestly oligarchy, who administered the economy, civil government and religious life (Allchin & Allchin, 1968). Also, there is a great deal of economic and political uniformity[3] in the vast region of the Indus Valley Civilisation.

On the contrary, there is a serious lack of study of urban history of any particular Indian city, post-Indus Civilisation, with very few exceptions[4] (Spodek, 1980). Unlike in the case of Indus Valley Civilisation, little is known about the second urbanisation in India. Even though detailed works have been carried out on the Indus Valley Civilisation, for which there are adequate archaeological evidences available, there is an acute paucity of information on later civilisations. Some authors have looked at literary sources to discuss the settlements of second urbanisation (see Amita Ray's *Villages, Towns and Secular Buildings in Ancient India*). In reference to reliance on archaeological evidences Ratnagar (2002) writes that the "orthodox barriers of archaeological method and subject-matter are now rapidly breaking down", but the area is still under development.

In 1974, R. S. Sharma wrote (as cited in Spodek, 1980, p. 252),

> If we leave out the problem of urbanisation in the Harappan sites, practically no work deals with urbanisation in ancient India. Some books catalogue and describe ancient towns chiefly on the basis of literary sources, but practically nothing has been done so far to examine the results of excavations in the last 25 years to explain the rise, growth, lay-out, etc., of the urban sites.

In 1984, archaeologist Makkhan Lal carried out horizontal excavations in the northern region that for the first time, and his book is still regarded as an important source of information on the study of settlement pattern in North India. Although the contents of the book are set on physical evidences of archaeological importance and may require further analysis on other issues, it still is a good source of information. In summary, the story of ancient cities is still incomplete, and continuous attempts are being made to knit the archaeological and textual evidences together. The following section puts together important points from the available literature and explains the nature of settlement during the second urbanisation.

In the light of the discussion on the inadequacy of archaeological evidences, the biggest limitation to Childe's (1950) approach is that it depends upon physical evidences and overlooks intangible characteristics of an urban society. Therefore, applying Childe's (1950) framework is difficult in the case of second urbanisation because there is inadequate archaeological evidence and reliance is upon literary sources of information which are only partially reliable. Erdosy (1985) strongly criticises Childe's (1950) approach and looks at intangible contributions of early civilisations. He redefines urban centres as "agents of social change" (Erdosy, 1985, p. 87) and claims that "the second flowering of civilisation on the Subcontinent was no less spectacular than the first (because) it was more durable as its effects are being felt to this day" (Erdosy, 1985, p. 82). Erdosy's approach is also supported by earlier theory of Redfield and Singer (1954) who define cities as the centre for exchange of intellectual ideas and physical goods. Later, Thapar (2002) reemphasises that exchange of ideas (and goods) was the nexus between different settlements within India and across the boundary (Thapar, 2002).

With the help of these discussions, this chapter widens the perspective of "city" to include even those that are found only in ancient texts like the 'Indraprastha'. Along similar lines of discussions, the next section explains 'urban form', as a physical design outcome and also as the result of social surgeries.

3.2 Defining 'urban form'

The form of a city is understood better when the determinants of its formation are explained. As mentioned in Chapter 1, the objective of this book is to develop a comprehensive understanding of Delhi, and therefore its urban form will be explained by understanding design and non-design factors that have shaped the city the way it is today. This section looks at the definition of urban form, as explained by design discipline scholars and also the determinants which guide human settlements, as explained by social scientists. At this stage, these discussions will open our thoughts towards non-design factors, which cause changes to the urban form. As and when we reach the conclusion, we will revisit these discussions to recreate a list of determinants of urban forms in the context of Delhi.

The term 'urban form' is at times used interchangeably with 'urban morphology', and both lack formal definition that can be shared across disciplines (Oliveira, 2016). In the simplest way, the two terms are differentiated and explained by Oliveira (2016): "urban morphology means the study of urban forms, and of the agents and processes responsible for their transformation, and that urban form refers to the main physical elements that structure and shape the city – urban tissues, streets (and squares), urban plots, buildings, to name the most important" (Oliveira, 2016, p. 2). This draws from the definition suggested by the Urban Morphology Research Group (1990) (referred by Oliveira, 2016) that defines urban morphology as "the study of the physical (or built) fabric of urban form, and the people and process shaping it". Moudon (1997) (referred by Oliveira, 2016) proposes a definition that includes historical evolution: "the study of the city as human habitat . . . Urban Morpholigists . . . analyse a city's evolution from its formative years to its subsequent transformations, identifying and dissecting its various components".

Study of urban form is usually understood as the study of physical outlook of the city. As mentioned earlier, urban form is often associated with design disciplines, particularly architecture, urban design and planning (Marshall & Caliskan, 2011). On the other hand, the study of the processes which form the 'urban' is defined as 'urban morphology' and is associated with the discipline of geography and spatial analysts (ibid). There is a continuous effort to integrate theoretical lessons acquired from studying the process of city building into the practical design of cities. The idea of merging findings from morphological studies into design is not new and is visible in works of Cristopher Alexander (1965) and John Frazer (1995). In his widely cited essay 'A city is not a tree', Alexander (1965) compares natural versus planned cities and confesses that attempts of artificially designing cities have been unsuccessful. Along similar lines, John Frazer (1995) uses technological tools to include elements of natural factors into the artificial creation of cities.

Frazer (1995) confesses the limitation of design theories and states that even though "the generation of form is fundamental to the creation of all natural and all designed artefacts. . . . There is no theory that can explain the creation of the urban form" (Frazer, 1995). To overcome this limitation, this chapter brings into discussion the discourse presented by the non-design-based spatial science of archaeology which provides a theoretical explanation to the evolution of cities and urban forms. Thus, there is a need to widen the lens of study of the built environment so as to include, into the cone of vision, the intangible elements like economy, society, polity, religion and other similar forces that act together to give shape to the built environment.

In the process of deriving the appropriate methodology for studying the historical settlement patterns created by human societies, archaeologists have scrutinised information from the fields of ethnology and geography and have constructed a set of determinants that affect the pattern of human settlements at various levels. In this context, a very comprehensive, path-breaking study was performed by Trigger (1968), who analysed literature from domains of ethnology, archaeology and geography and concluded that settlement patterns can be grouped into three broad categories (Trigger, 1968): "(i) individual building or structure; (ii) the manner in which these buildings or structures are arranged within a single community; and (iii) the manner in which communities are distributed over the landscape" (Lal, 1984, p. 164), similar to the built environment components discussed in Chapter 1.

On another note, a similar hierarchy of the urban form is discussed by Kropf (1996), although in reference to the field of architecture. To make easy the complexity of urban form, Kropf (1996) first explains "complexity" and then suggests a framework to decode the complex urban forms. He says that

> complexity is an essential concept for an understanding of towns. It is central to the notion of form, and lies at the root of any theory of built form. . . . To give a basic definition of complexity, an object (or city) is complex when it is judged to be composed of several smaller objects. . . . When we recognize something as complex, we recognize whole objects and the parts of which they are composed and, by implication, a relation of part-to-whole . . . in general, form can be defined as an arrangement of parts considered as a whole.
>
> (Kropf, 1996, p. 251)

Since the city is a complex entity, each step, from part to whole, would mean a step up in level of complexity. Each part is composed by sub-parts, and in abstract sense, this structure continues to infinity. By convention, the building materials can be taken as the smallest unit in this series, which has no sub-parts (Kropf, 1996). With the help of the part-to-whole concept, Kropf (1996) establishes a formal definition of the otherwise loosely defined term of "urban tissue". He writes that "from the viewpoint of hierarchy, urban tissue is, in effect, a synthesis of all the components. It is an organic whole that can be seen at distinct levels of resolution". (p. 252). At a lower level of resolution, a 'tissue' may be described as an arrangement of streets and blocks. The next detailed level will be to look at a plot, in a series of plots. At a higher level of resolution, the details of the buildings and materials may be seen.

The formats suggested by Trigger (1968) and Kropf (1996) are used by archaeologists and architects, respectively, and possess similarity in terms of various levels of resolution of the cityscape. That said, the archaeologists further discuss different sets of factors that affect the settlement pattern at each level. Trigger (1978) writes that

> if we conceive of the settlement pattern as an outcome of the adjustments a society makes to a series of determinants that vary both in importance and in the kinds of demand they make on the society, we must consider not merely the range of factors affecting the settlement patterns but also the manner in which different factors interact with one another to produce a particular pattern.
>
> (p. 189)

As a caution, it is important to mention that the determinants found by Trigger (1968) are in the context of studying archaeological remains of ancient cities, and therefore the findings may not be directly applicable and need to be tested in the case of contemporary cities.

3.2.1 *Explaining the determinants at each level of resolution*

This section extrapolates extensively from Trigger's (1968) chapter "The determinants of settlement patterns". At the first level, the factors affecting an individual building are discussed. Following are the take-away points:

Level I – individual building

Uses of buildings:

a. Trigger (1968) finds that special purpose structures are more common in class-divided societies. He explains that higher the level of complexity of the society, higher is the specificity of use of buildings, for example house, temple, tombs, forts. Religious beliefs of the society may also determine the construction of shrines, temples or tombs. Also, there are a variety of public buildings in complex societies that are devoted to secular group activities, like schools and libraries, public bathrooms, stadia.
b. Further to that, more complex political organisation will demand buildings that serve the special needs of the administration for example, offices, jails and barracks.

Building materials:

c. In the context of historic cities, the choice of building materials was made based on the range of materials available in the location and also considering the performance of the material in protecting against the climate.
d. An interesting determinant of building materials is the subsistence regime of the society. This means that depending upon whether the occupants are migratory or permanent settlers, the type of building structure may be permanent or temporary. Temporary structures can either be erected at the place of camping if natural materials are available on site. Alternatively, the temporary structures may be designed to be portable and carried along to locations where there is a scarcity of natural materials, like in desert areas.

Structural design:

e. The structural design of buildings is reflective of the skill and technological know-how of its builders. Their level of understanding of performance characteristics of different building materials will also influence the structure of the building. For example, the use of brick and stone in Europe and Western Asia allowed erection of large elaborate structures with load bearing walls. In the wood architecture of East Asia, poles and beams provided a structural framework on which the roof and walls were hinged.
f. The climatic conditions (particularly the temperature, humidity and air movement) of the place are another important factor that impacts upon the structural design. For example, in desert climates the diurnal range is high, and, therefore, thick walls of stone and clay are used because they will absorb heat during the day and release it night.

Size, shape and layout of houses:

g. In societies where social hierarchies, based on class, caste, wealth or authority, are more pronounced, the size and grandeur of the house also respond to the stature of its resident. This means that the house of elite will be relatively larger and more elaborate.
h. The size and layout of a residence directly responds to the size and structure of the family. The house for nuclear family is small, with one or two rooms, and lineage families had large houses with multiple rooms. The organisation of a family may vary, depending upon social practices of having residential servants or multiple partners.
i. The shape of the house is also guided by the climatic conditions. For example, in warm climate zones, houses are built around an open courtyard, whereas in cold areas, the houses are usually compact (e.g. an igloo) so as to ease heating. The climate also influences the orientation of buildings to a huge extent.

j. Lal (1984) refers to findings of ethnographers Robbins and Flannery that suggest the correlation between shape of the house and substince pattern. The findings suggest that circular and rectangular houses were created by temporary and permanent settlers. Archaeological studies also suggest a shift from circular to rectangular dwellings, following the change in subsistance pattern. In cases when the household specialises in home-based economic activities, the house has a dedicated place as a workshop or storehouse.
k. The maturity and quality of political systems may also influence the layout of the house. For example, if the security system is weak and incidences of crime and attacks are common, then houses are built in the elements of security. This may mean inward-looking houses or high peripheral walls which are strong enough to hold off attacks.
l. Cosmological conceptions of a society may also influence the orientation and layout of a house, as also observed in Hindu city planning.

Level II – community

Trigger (1968) used Murdock's (1949) definition of community, which is, "the maximal group of persons who normally reside in face-to-face association" (p. 60) and later modified the definition to include "the whole large units of settlements, such as cities, which, if they cannot be defined as communities, at least represent stable interaction patterns" (p. 61). This widens the scope of discussion to include different levels of settlements, such as neighbourhoods, villages and cities.

Size and location:

a. People tend to settle at locations that are close to drinking water, sources of food, and are safe and pleasant to a satisfactory level. These ecological factors are found to be the major determinants of community size and locational stability.
b. More complex societies could improve subsistence technology and develop trade and commerce of surplus, as discussed in Chapter 1. Trade provided the source of income which then can be used for the development of agriculture in areas where otherwise it is difficult to cultivate food, or else income is used to import food from other locations. Settlements may then develop in any location, including wastelands rich in mineral resources, like gold, silver and copper, but poor in agricultural productivity.
c. Strength of family ties may also determine the pattern of settlement and relocation behaviour of a community (Chang, 1962). For example, referring to circumpolar societies, Chang (1962) writes that in Siberian types of communities, the family (from descent or from marriage) is the "basic unit of economic cooperation" and is strongly bonded together, thus preferring to settle or move together. On the contrary, the Eskimo types of communities

are incoherent conglomeration of families that are loosely bonded, and each family may settle wherever it likes and move to a new district offering advantages.

Layout:

d. The layout of the community was influenced by family and kinship organisations, and each lineage will usually occupy a section in the village. Lal (1984) writes that in more complex societies divided by social class, caste, religion and ethnic groups, the layout may demonstrate a spatial segmentation of these social groups. As mentioned in Chapter 1, the specialisation of trade also encouraged artisans associated with a particular craft to live in the vicinity so as to facilitate the obtaining of raw material and selling of the finished goods to the merchants (Thapar, 2002). This can link back to kinship organisations if the trade is inherited, as in the case of caste system in India. Correlation between social and spatial organisation is not entirely confined to primitive societies and has continued till date, in rural settlements in India.
e. Public buildings which serve as the loci of all social activities are usually located in the centre of the city, thus making them equally accessible for all the members of the community. Also, at times the houses of the elite members are centrally located, although this may not always be the case, as observed in the cities of Indus Valley Civilisation where the elite residences and citadels were at one side of the city.
f. Level of complexity of political organisations and defence system may also influence the layout of a community. For example, the Roman Empire once had an adequate army to safeguard its cities that were sprawling to form a new agglomeration. However, the defence system collapsed in the third century AD, and the population was shifted from suburbs into the crowded walled cities. At times, isolated forts and garrison towns were built to safeguard the countryside (Trigger, 1968).
g. Unstable political environment, political tensions and warfare also play an important role in determining the location of a settlement. When wars are frequent, the settlements are located on hilltops or in bends of rivers. Also, the central area of the city or major buildings are walled or fortified to serve as the place of last retreat.
h. Similar to house layout, the city layout and orientation are also guided by cosmological beliefs, as observed in Hindu city planning.

Level III – zone (or region)

Ecology:

The population density of a region is influenced by the availability of natural resources that are being exploited. Settlements are avoided on lands which are poor in natural resources or are prone to diseases or other dangers.

Fertile regions become the centres of population, thus gaining political and cultural importance. Trigger (1968) gives the example of Japan, where the main fertile plains of Kanto, Nobi and Kinai host the chief cities and for which there has always been a struggle for control.

Demography:

In addition to that, there may be sudden rise or fall of population consequential to severe ecological changes or economic and political changes. This may tend to influence the settlement pattern to a great extent. There could be other reasons for a sudden change in population, like the bubonic plague in Central Europe that resulted in sudden depopulation, thus converting many farmlands into forests between 1350 and 1450 AD.

Political structure:

For security reasons, there exist various administrative towns and garrisons in various sections of the country. As mentioned earlier, the hunting and food gathering communities had a strong sense of territory, which they safeguarded. However, there was no immovable property, and therefore they moved out to another region in case of danger of war with others. However, the pastoralists developed another way to safeguard their land and property, by which they would come together to form a bigger group at the time of war. Also, they would join together to build forts where they could flee at the time of danger. Depending upon the level of political organisation, the security measures would be different. For example, at a village level, the fort was the headquarters of the village chief; where there were a number of villages, the largest village was fortified; and in the case of city states, the capital city served this purpose. In large states, the defence system may become an important element of the settlement pattern, also demanding extensive road and transport systems for the movement of troops and messengers.

Religion

Religious factors may also influence the settlement pattern of a region. Trigger (1968) gives example of Judeo-Christian religious communities of Middle East who aspired to escape the world and settled in isolated and lonely regions which were otherwise difficult to develop.

Economy:

Zonal patterns are influenced by economic factors to a great extent. In particular, trade plays an important role in the growth of population and in the creation of new cities or the expansion of existing ones. Trigger (1968) explains this taking the example of modern industrial cities which neither produce food enough to support their own population or the raw material required for its industries, and both are procured through trade.

These determinants have been useful for studying the historic evolution of cities, and in understanding the pattern of settlements at different scales, starting from building to zonal pattern. These factors are spatial elaboration of the built environment components discussed in Chapter 1 in conjunction with the institutional factors that shape the built environment, also discussed in Chapter 1. The preceding factors may also be believed to influence the future of urban settlement patterns of contemporary cities, including Delhi. Kropf (1996) writes that "the physical structure of any town is a record or documentation of its creation" (p. 254) and that

> the process of formation of any city is history, regardless of the short distance in time. . . . What differentiates the past from the present are those characteristic features that bring up the idea of distance in time and make visible the change in characteristics over time.
>
> (Kropf, 1996, pp. 254–255)

Therefore, this framework used by historians shall apply well in the case of modern cities, particularly Delhi, the form of which has been changing due to various factors. Chapter 1 tried to explain the social, economic and political elements. The above archaeological framework will help in understanding the physical form of each layer of settlement that ever existed in Delhi. Put another way, the journey of Delhi, from being a modest village settlement in ancient time to the capital city-state of modern India, will demand discussions on different scales at which the city has ever operated, and, therefore, the determinants of its urban form, at each stage, may also change. Adopting an archaeological framework will make possible the comparative study of the multiple layers of settlements that top each other in Delhi.

Certainly, these determinants are contextual to historical cities and may not be exactly applicable on contemporary urban settlements. Therefore, appropriate modifications and additions may be made, and new factors may also come into existence as and when we progress in time to discuss the urban form of modern Delhi.

Taking guidance from the aforementioned discussions, this book defines urban form as the dynamic physical appearance of the city that is consequential to organic and planned response to the needs of the society (social, economic, political, cultural, religious, and any other). Urban form is the outer line of physical positions, of societal functions, on the existing natural environment.

3.3 Delhi as a part of the first urbanisation in India – Indus Valley Civilisation

Between 2800 BC and 1500 BC,[5] the Indus Valley Civilisation emerged as the 'first urbanisation' or 'Bronze Age urbanisation'[6] of early India (Thapar, Ancient Indian social history: some interpretations, 1987). It expanded over a vast geographical area around the Indus River Valley (in Sindh, Pakistan), extending from the Himalayas and Hindu Kush in the north to the coastal regions of Kutch and Gujarat (India) in the south; western extension up to Makran coast in Baluchistan and eastward, touching Meerut (India) and the other western fringes of Uttar Pradesh (India), covering an area larger than that of Mesopotamia or of Egypt (Good, Kenoyer, & Meadow, 2009).

Evidences found in Mehrgarh (from 6000 BC) suggest the connection of the Indus Civilisation with Central Asia, Iran and Afghanistan. Ratnagar (2002) calls them the "wild ancestors" of this Bronze Age civilisation who were discontinuously distributed through the uplands of Asia from Turkey to Afghanistan (p. 4). The transition from being hunters and gatherers (or Mesolithic) to agriculture and animal-rearing society (or Neolithic) was happening in different parts of South Asia at different times (Ratnagar, 2002), although each region had its unique stone and ceramic technology and a range of domesticates (ibid). Ratnagar (2002) argues that the geographical location of a developed farming settlement is not simply determined by any specific environmental character and rather is dependent on many technological and historical factors; however, they do not explain these factors in detail. These other factors may be linked back to Trigger's (1968) list of determinants discussed earlier. Trigger concludes that the regions with good soil quality and high population density are not necessarily the ones which developed farming economies (Ratnagar, 2002). There is a dearth of evidence on the early pattern of land use in South Asia, and this limits the discussions on the topic.

3.4 Delhi as the sacred city of 'Indraprastha' – Vedic period

The roots of this "city" are probably 5,000 years old, and a lot still remains to be explored. Delhi holds some remains of the first urbanisation in India (Indus Valley Civilisation between 2500 BC to 1550 BC (Basham, 1969); at the time, it was probably a modest settlement at the eastern outskirts. Excavations were undertaken by the Archeological Survey of India in various phases, and Painted Grey Ware was found at places in and around Delhi. However, this does not confirm the spread of the Indus Civilisation at this location. Later, the stories of the great Indian epic 'Mahabharata' discuss 'Indraprastha' as the capital city of Pandavas during the second wave of urbanisation (or the Vedic period between 1500 BC to 500 BC, Basham, 1969), thus bringing 'Delhi' into the limelight as a capital city for the first time. As per the inscription of Chandravati's 'Gahadavala', c. 1091–1103 A.D., the king was expected to protect the scared places, which included 'Indrasthaniyaka' (or Indraprastha), Kasi, Kusik, and Uttara-Kosala (ibid). The mention of Indraprastha in literary sources puts an emphasis on dating back 'Delhi' to be the capital city of Pandavas. Although the textual corpus of the great epic 'Mahabharata' describes the city of 'Indraprastha' in detail, there is a dearth of archaeological evidences which support these descriptions. In fact, the date of these descriptions is not established yet (probable date being 1450 B.C., Hearn, 1997), and the uncertainty increases further because the epic was written much later than the time of occurrence, and the date of writing of the epic is itself doubtful (Thapar, 2002) (probably 'Mahabharata' was written around the fourth century AD; Ray, 1964).

In the 'Adi Parva', the following elements of Indraprastha are mentioned (Ray, 1964) (Figure 3.1):

1. City was well fortified on all sides;
2. Wide ditches circumscribing the city;
3. Entry gates in all four directions;
4. Palatial buildings inside the city;
5. Towers inside the city (probably watch towers near entry gate);
6. Long and wide thoroughfares dividing the city into various sections or squares;
7. Magnificent houses, pleasant retreats;
8. A number of small and large tanks brimming with water;
9. Beautiful lakes fragrant with lilies and lotuses;
10. Large parks with pools at the centre, artificial hills and so on.

In the context of the name of the city of Delhi, Beglar (1874) writes that "it is the universal custom to name cities after their founders", and Delhi is not an exception. The Vishnu Purana mentions Raja Dilipa as having preceded

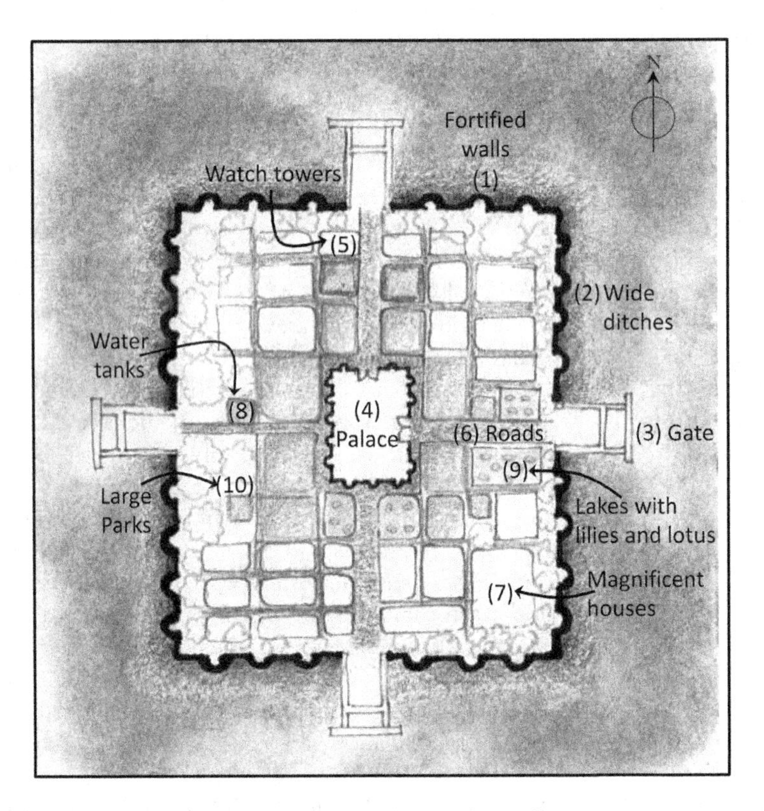

Figure 3.1 Layout of city of 'Indraprastha' as described in 'Adi Parva'

Source: Authors

Pandus (ibid) of Mahabharata. However, Indraprastha overshadowed Dilipa's "Dilli" for some time, and Dilli could regain prominence only after the decay of Indraprastha (ibid). There are presumptive evidences that support the theory of decay of 'Indraprastha' and the rise of 'Dilli' as a place of importance (ibid).

Moreover, Delhi remained unmentioned, after the decay of Indraprastha, for a long time of 800 years until Angapal rehabilitated the city and constructed the fort of Lal Kot around 1052 AD (as will be discussed later). It is hard to believe that the city was totally deserted for this time period, and Beglar (1874) suggests that "although Delhi was not the residence of the paramount king, was nevertheless in existence and flourishing in a humbler way", and the rulers of Delhi were probably acknowledging the supremacy of the raja of Ujjain in the east of India. Based on this theory of Beglar's (1874), we assume that the nature of the settlements in Delhi during the later Vedic period would have been similar to that of other ordinary cities.

3.5 Delhi rehabilitated as Lal Kot (1052 AD)

Lal Kot presents the oldest remains of identifiable building structures in Delhi that are "substantial and extensive enough to be called a city" (Blake, 1991, p. 9). As discussed earlier, after the disappearance of Indraprastha, there was a long gap of around 800 years before Delhi was re-peopled and the fort of Lal Kot was constructed by the Rajput Tomar ruler Angapal (Beglar, 1874). Although the date for the city is not known with certainty, Beglar (1874) estimates it to be built around 1052 AD.

Cohen (1989) refers to the *Pasanahacariu* (the hagiography of the twenty-third Jain 'Tirthankara') and explains the political environment of North India in the end of the tenth century AD. He writes that the Tomaras were not powerful enough at that time to compete with the powerful regional dynasties of the Gahadavala at Kannauj-Varanasi and the Chauhana in Rajasthan and were annexed to either of these two kingdoms as feudatories or else ruled with their consent (ibid). A little over a century later, the Tomars were defeated by the Chauhans who were just another clan of Rajput rulers (Blake, 1991). Around 1180 AD, the Chauhan ruler Prithviraj, or Rai Pithora, founded a new fortified city by expanding upon Lal Kot, called Qila Rai Pithora (ibid). This was mainly in response to the increasing attacks from Afghans in the northwest (ibid). Multiple raids were performed over North India by Mahmud of Ghazanavi and later by Muhammad Ghori focusing on northwestern territories of Chalukya, Chauhana and Gahadavala (Cohen, 1989). Delhi was not reached until these central powers were neutralised. Finally, in 1192 AD, Prithviraj lost to Ghori, and Delhi was later captured by his Turkish slave Quṭb al-Dīn Aibak in 1193. It was a practical political decision of Muhammad Ghori and his slave Quṭb al-Dīn Aibak to locate their capital at Delhi given its central location between their Hindu territories and their homeland in Afghanistan (Cohen, 1989). While keeping their headquarters in Qila Rai Pithora, these early Muslim rulers in India were confining their building activities to reconstructions and renovations. The mosque of Quwwat-ul Islam (the power of Islam) and the famous minaret called the 'Qutub Minar' are two important architectural symbols of this time (refer to Chapter 4 for the architecture and material details of the two structures).

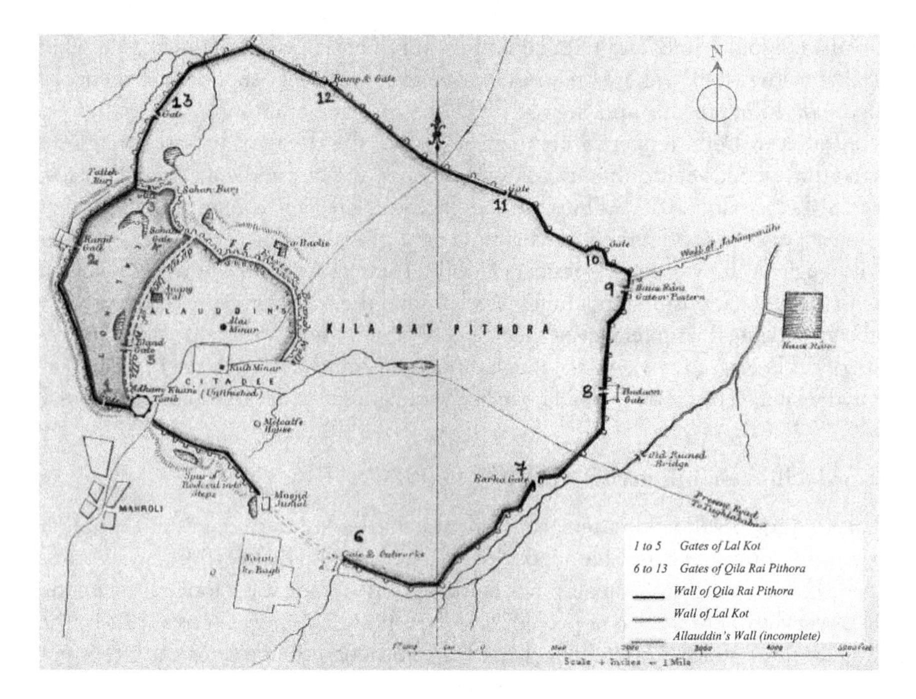

Figure 3.2 Lal Kot and Qila Rai Pithora

Source: Authors (original image sourced from Beglar 1874, Plate I)

Without getting into the details of political environment, which are explained in the Chapter 1, we will rather focus on important urban centres that were created during the Muslim rule in Delhi. Together with the Hindu city of Prithviraj Chauhan, these are popularly addressed as the "seven cities of Delhi", as mentioned next:

City	*Name of the city*	*Built by*	*Year of construction*
1st	Lal Kot and Qila Rai Pithora	Anangpal and Prithviraj Chauhan (1170–92), respectively	1052 1180
2nd	Siri	Alauddin Khilji (1296–1316)	1303
3rd	Tughlaqabad	Ghiyath al-Din Tughlaq (1321–25)	1321
4th	Jahanpanah	Muhammad bin Tughluq (1325–51)	1325
5th	Firozabad	Firuz Shah Tughluq (1351–88)	1354
6th	Dinpanah and Shergarh	Humayun (1530–55) and Sher Shah Sur (1540–45), respectively	1533 1540
7th	Shahjahanabad	Shah Jahan (1628–58)	1639

Source: Authors; data sourced from Blake (1991)

The unstable political environment was directly translating into the built form, and we observe frequently emerging capital cities in Delhi, each created by a new ruler from a different dynasty. This continued to happen until the Mughal dynasty was established by Babur in 1526. However, the Mughals were facing many challenges to establish control over North India; in the process, the capital was shifted between Agra, Delhi and Lahore. Sinopoli (1994) marks the number of capital cities that were built by the Mughals and writes that "the most striking feature of Mughal capitals is how many there were" (Sinopoli, 1994, p. 294). Sinopoli (1994) provides three probable justifications for the new capital city formations by the Mughals: (1) Cosmological beliefs: the Mughals believed that capital city is the cosmic centre ('axis mundi') and forms an essential component of the system of authority of the ruler; (2) "Mughals had exceptionally wide interests or scope in comparison with other cities", and they viewed the capital city as a symbol of government ideologies, political strength and racial superiority; and (3) political control and efficient administration: the location of the capital city plays a "primary role in the organisation of territory" (p. 296).

One of the most noticeable Mughal capital cities in Delhi was Shahjahanabad, which was built by Shah Jahan in 1639. The distinguishing feature of this city is that it has outlived the empire that built it many centuries ago, to be operational as "Old Delhi" of modern times. This has recently caught attention of researchers interested in studying the city's evolving urban character and the challenges it faces in serving the growing needs, which will be discussed later in the following sections.

3.6 Shahjahanabad – the imperial capital of the Mughals (1639)

The new imperial city was built between 1639 and 1648 and remained the home for the Mughal emperors until 1858 (Blake, 1991). The cityscape of Shahjahanabad, like the other sovereign cities of Istanbul, Isfahan, Edo and Peking, is dominated by palaces and mansions or other places of importance for the emperor and his nobles (ibid). Taking forward the debate on the definition of cities, Blake (1991) strongly contradicts Max Weber's notion that the cities in Mughal India were merely princely camps, and puts forward a strong argument that these cities were a miniature empire in itself. The sovereign cities like Shahjahanabad were "urban communities" with their own distinctive style and character, dependent on "a particular kind of state organisation" (Blake, 1991, p. xv).

As explained earlier, the creation of new capital was often driven by cosmological importance of capital city that was prevalent among Mughal rulers. As a probable justification for the creation of Shahjahanabad, Blake (1986) writes that

> [a] capital stood as a symbol of the ruler's power and wealth, an example of his ability to order the world about him into regular, harmonious, even beautiful shapes and patterns. In civilisations such as those of Mughals, where the capital was the "axis mundi" – the centre of the earth and the intersect of the

> celestial and the mundane – the need to choose an appropriate site was even more acute. . . . As a choice for axis mundi, this site (on a bluff overlooking the River Yamuna) was hard to fault.
>
> (Blake, 1986, p. 153)

The physical layout of the city of Shahjahanabad is influenced by Hindu and Islamic ideas of city planning (Blake, 1986). The mixed character of the plan of the city is a good reflection on the demographic mix of this city which was predominantly a Hindu population governed by Muslims and also the capital city of Muslim dynasty in a Hindu subcontinent (Blake, 1991). The shape of the city is semi-elliptical which is similar to *karmuka* layout, as mentioned in the Hindu vastu shashtra 'Manasara' (Blake, 1986). The 'Mansara' discusses multiple types of layouts for villages of which *karmuka* (meaning bow-shaped) layout is specially recommended for settlements on the side of a river or sea (Acharya, 1980). Shahjahanabad is in close proximity to the River Yamuna, and, therefore, the bow shape is recommended by 'Mansara'. In reference to the functioning of the city, the *karmuka* layout was recommended for trader's settlements or for the 'Vaishya' caste (Acharya, 1980) probably to make better use of opportunities for long-distance trade due to closeness to water bodies. Regarding the function of Shahjahanabad city, the discussions are usually around administrative roles of the city as the capital of Mughal sultanate, and it is less known if trade and commerce were among other important functions of the city. This raises doubts that even though there is a specific layout recommended for capital city (or *rajdhani*) in 'Mansara', why was Shahjahanabad laid out as *karmuka*? The answer is not clearly known and will require further investigation. A plausible explanation could be that the semi-elliptical shape was found visually impressive and functionally suitable for the topography around the River Yamuna. On another note, the practical application of Hindu plans is itself doubtful, and Ray (1964) suspects that even though various types of city plans, based on its dominant role, for example, administrative town, port town, market town, are mentioned in 'Arthashashtra',[7] the idea of city plan was more of "a design in abstraction" rather than a "social reality" (p. 51).

While the shape of the city reflected Hindu style of city planning, the internal layout was inspired from the Persian planning (Blake, 1986), in which the organisation of spaces inside the city are in line with human anatomy and vice versa, which means that the human body is constructed like a city (Blake, 1991). The cosmological doctrines of Persian planning are applied to Shahjahanabad city, as explained here:

- Shahjahanabad is a walled city, which symbolises the cosmos. It has gates and towers all around, most of which remain today. These gates are believed to be the entry into four cardinal directions plus the four doors to heaven (Blake, 1991).
- The Chandni Chowk is the central bazaar and the main collector street, similar to the backbone of the human body (ibid).

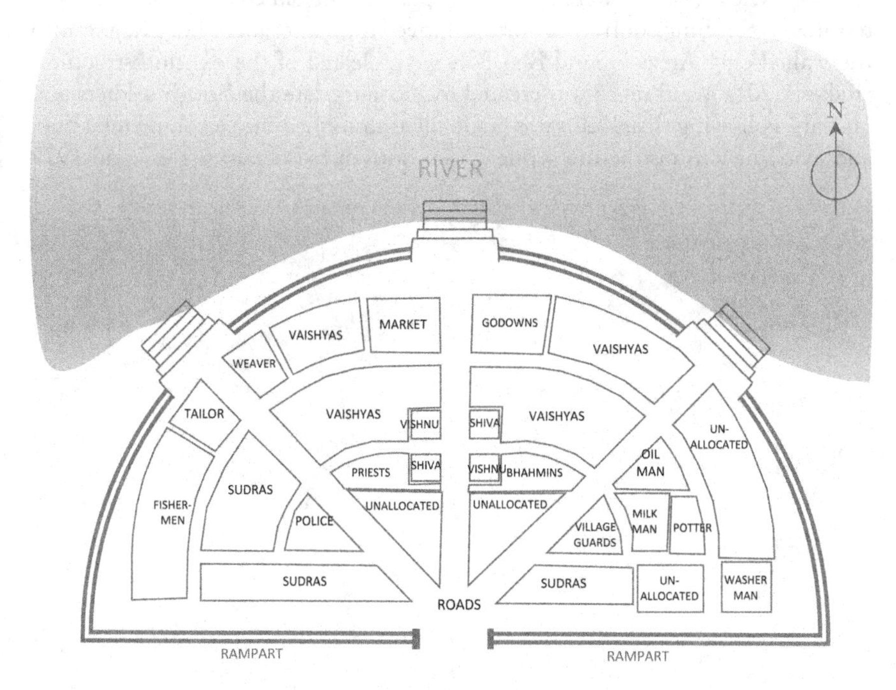

Figure 3.3 Plan of *karmuka* village as described in 'Mansara'

Source: Authors; original image sourced from Acharya (1980, Plate XXI)

- The street starts from the palace fortress, the head, and runs west to cross the Jama Masjid, or the heart of the city, to reach the Lahori Gate (ibid).
- The orientation of the city was such that the palace fortress faces the west, where lies the scared city of Mecca (from India); Jama Masjid opens up in the west, and also the Chandni Chowk runs towards the west (ibid).

3.7 Transformation of Shahjahanabad into a commercial city – 1857 to 1931

After the mutiny of 1857, the British cautiously established their control over all major cities in India, particularly Delhi, which was the seat of the Mughal Empire. "There was a need to diminish the perceived grandeur of the Mughals in the eyes of the natives" (Khosla & Rai, 2005b, p. 14), and several colonial interventions were made in Shahjahanabad (Mehra, 2013) with intentions to establish a better control over Delhi, and over native Indians in general.

Figure 3.4 shows (1) the British cantonment near Northern Ridge; (2) the pontoon bridge crossing the River Yamuna; (3) old city walls which basically contained the entire city population; (4) major gates; (5) the Red Fort of the Mughal emperor; and (6) other elements of the city like bazaars, *baghs* or gardens, residences of 'nawabs' and government officials.

The physical changes were about to happen to the old city. As an initial step towards establishing control, the British moved the cantonment from Rajpur into the walled city. Areas around Red Fort were cleared of the existing structures (Rakesh, 2014), and space was created to accommodate the British soldiers and officials. Following this, Delhi was gradually transformed into a commercial hub, and many new infrastructure services were introduced between 1867 and 1923

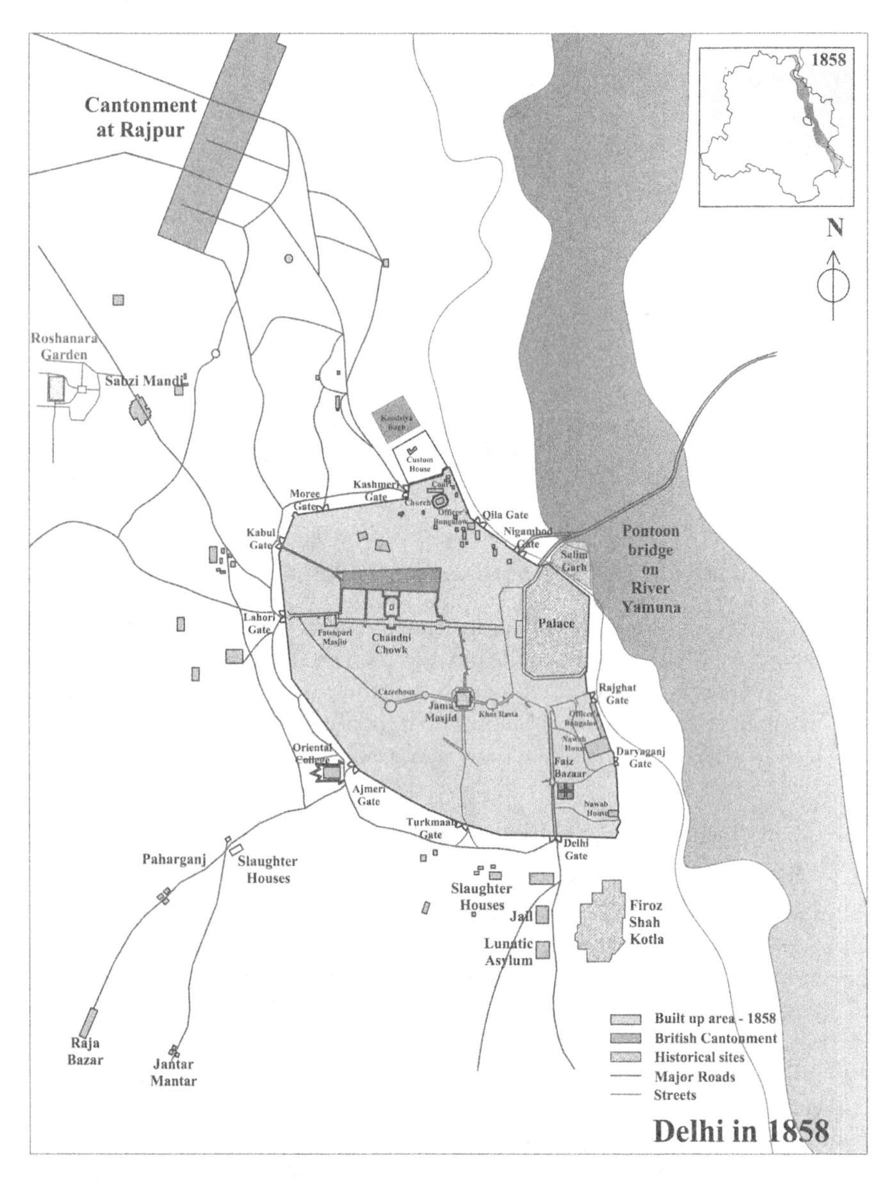

Figure 3.4 Map of Delhi and its surrounds in 1858, before its transformation into a commercial centre

Source: Authors; original image sourced from Weller (1858)

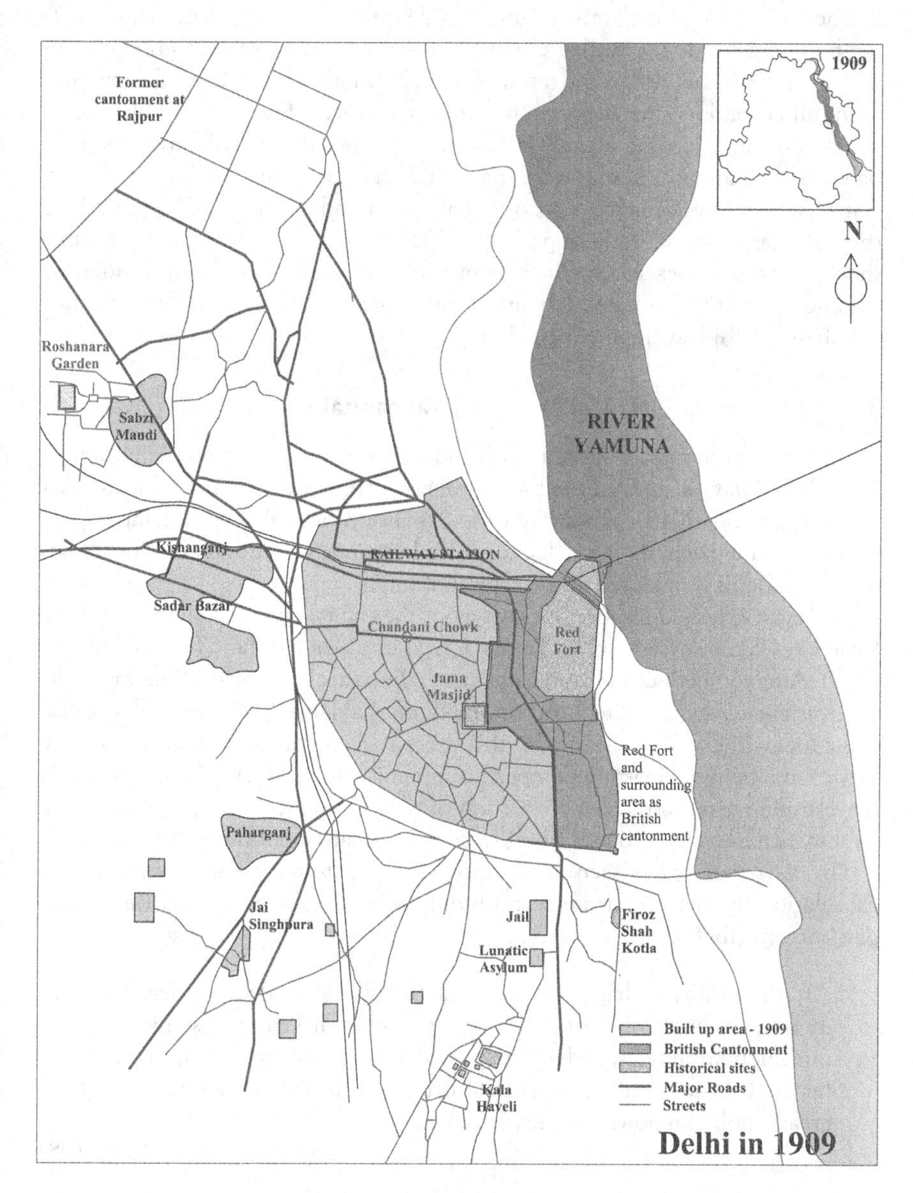

Figure 3.5 Map of Delhi and its surrounds in 1909, after it was developed as a commercial hub

Source: Authors; original image sourced from *Imperial Gazetteer of India* (1909)

(Mehra, 2013). These included a railway junction (1867), waterworks (1892), electric trams (1901–02), electricity (1902), drainage (1909) and telephone lines (1923), a few of which still support the city (ibid). Development of commerce and infrastructure boosted population growth, and an increase of 41% was witnessed in this zone in the decade of 1921–31 (ibid).

The population growth raised the demand for space; both residential and existing buildings inside the walled city were extended and stratified to house commercial activities as well as the resident population. However, the walled city was soon full beyond its carrying capacity and could not confine these developments in its boundaries. Being already in commercial use, the neighbouring areas of Sadar Bazar and Sabzi Mandi were the initial zones of expansion. Later commercial activities spread further west over Paharganj and Karol Bagh. Even to date, this zone (special area zone as per MPD 2021) hosts many formal and informal shops, mills, factories and small-scale manufacturing units as well as residential colonies of factory workers, labourers and similar other blue-collar workforce members (Delhi Development Authority, 2006).

3.8 'New' Delhi (1911–31) – a federal capital

Delhi became the capital of the British India Empire in 1911 when the unrests in Bengal, in a way, compelled the British to move out of Kolkata and relocate elsewhere (Johnson, 2015). Since then the expansion of Delhi, both demographic and spatial, has been alarming due to multiple events, before and after independence that Delhi shouldered as the capital of India.

The most expected location for laying out the capital city was the then existing location of European settlements around Civil Lines near Northern Ridge (Johnson, 2015). After lots of discussions and arguments the final choice of location was made, and this was in the opposite direction, south of Shahjahanabad, where New Delhi exists today (ibid). The selection of this site was even criticised for separating the new capital from the existing European community in Delhi Civil Lines. However, the planning team justified the rejection of the Civil Lines area on geographical, sanitary and historical grounds (ibid) (read more details in Johnson, 2015, p. 87).

The new capital was seen as an important step towards harmonising political relationship between Britain and India and as a stronger base for the British government (ibid).

> After Britain's problems in Bengal, the British Raj needed to redefine itself as incontestably strong yet benevolent enough to make concessions to deserving colonial subjects. Federalism provided an attractive hegemony that better unified the diverse provinces of British-India and moved Indians toward greater political power in the provinces.
>
> (Johnson, 2015, p. 108)

The concept of federalism was directly applied to the new capital, which was physically kept detached from any one province and was rather planned as a federal capital city. In doing this, huge inspirations were drawn from the then existing federal capital cities, particularly Washington, D.C. and Canberra.

The physical form of the capital city was to serve to the contradictory purposes of showcasing British supremacy and harmonising their relationship with native Indians. On one hand, the British were aspiring for a better relationship with the natives through the construction of a new, inclusive capital city "for all class and races; with their churches and temples, hospitals and schools" (Johnson, 2015, p. 92). On the

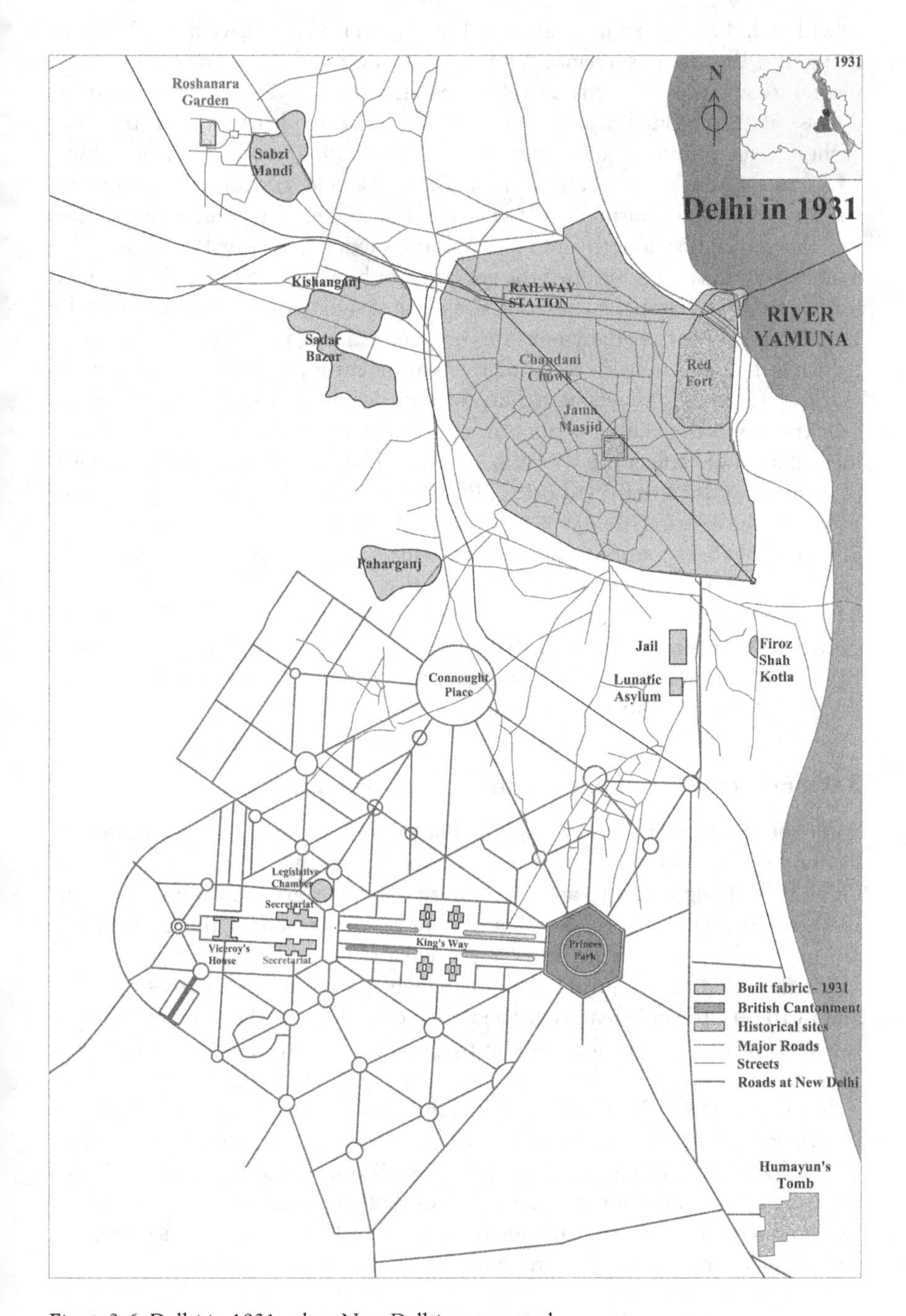

Figure 3.6 Delhi in 1931, when New Delhi was created

Source: Authors; original image sourced from www.columbia.edu/itc/mealac/pritchett/00maplinks/modern/delhimaps/britannica1910.jpg, Retrieved 12 May 2017

other hand, the personal inclination of Lutyen and leading Englishmen was to gift themselves the idea of supremacy and power of the British government. The town planning committee selected Edwin Lutyens, along with Herbert Baker, for this task.

"Just as the colonial layout of New Delhi was intended to demonstrate the higher urban culture of a superior Western civilisation, so Shahjahanabad was created to present to the people of India the higher and more ordered urban civilisation of Timurid Central Asia" (Khosla & Rai, 2005b). Thus, alongside nurturing Western civilisation, the new city was also subduing the Mughal, although in a subtle way. Lutyens positioned the government house on Raisina Hill and raised it higher than the minarets of Jama Masjid, thus signifying the supremacy of the British capital over Shahjahanabad (Khosla & Rai, 2005a). The idea of contextualising the new city with the Indian environment was proposed to Lutyens by Patrick Geddes. However, both Lutyens and viceroy of India, Lord Hardinge, "saw the city as the symbol of not only colonial power, but also of a new environment that would reflect the Fabian concerns of Ebenezer Howard about the ideal place for the ideal citizen" (Khosla & Rai, 2005a, p. 51). Despite Lutyens's insensitivity to the existing Islamic fabric of Delhi, the historic Mughal structures were placed at the focal points and at the ends of major vistas, which gained enormous popularity for his plan.

The new city was inaugurated in February 1931 (Johnson, 2015). The country gained independence soon after that in 1947, thus causing the departure of the British from an underutilised capital that was created, in the first place, to help them stay longer and stronger in India.

3.9 Fragmented growth of Delhi (1931 onwards)

After the creation of the new capital city, Delhi was spatially divided into two distinct urban centres which varied drastically in their character and built form. Even though the two areas were notionally one city, they were ways apart in their physical constitution. The differential approach of the government (Kishore, 2015) towards these areas was clearly reflecting in its built form. By 1922, the building footprint in Shahjahanabad (see Figure 3.7) was covering almost the entire land area inside the walled city, also sprawling over neighbouring areas, while New Delhi was scantly populated, with beautiful gardens and wide boulevards.

In the year 1942, Delhi became the permanent capital for all seasons, and the earlier practice of shifting the capital to Shimla in summer was ceased (Mehra, 2013). This was also the period of the Second World War, when India and Britain were partnering against Japan. The war led to a huge temporary inflow of foreign forces in Delhi. Spatial fragmentation of Delhi was easily noticeable, and a visiting American soldier wrote that

> [b]efore the (second world) war, there have been eight Delhis, six of which crumbled away with declining civilisation. [The] beginning of the war found still in existence a seventh Delhi, a typical teeming eastern city of rickshaws, sleeping coolies and wandering cows; and the eighth Delhi, a governmental

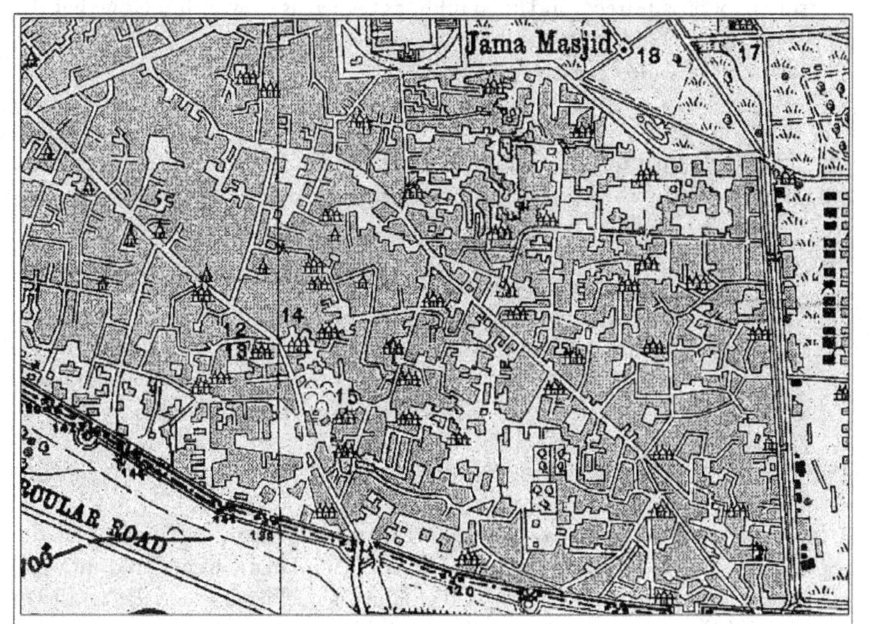

Built fabric in Old Delhi (up) and New Delhi (down) in 1922 (image in same scale)

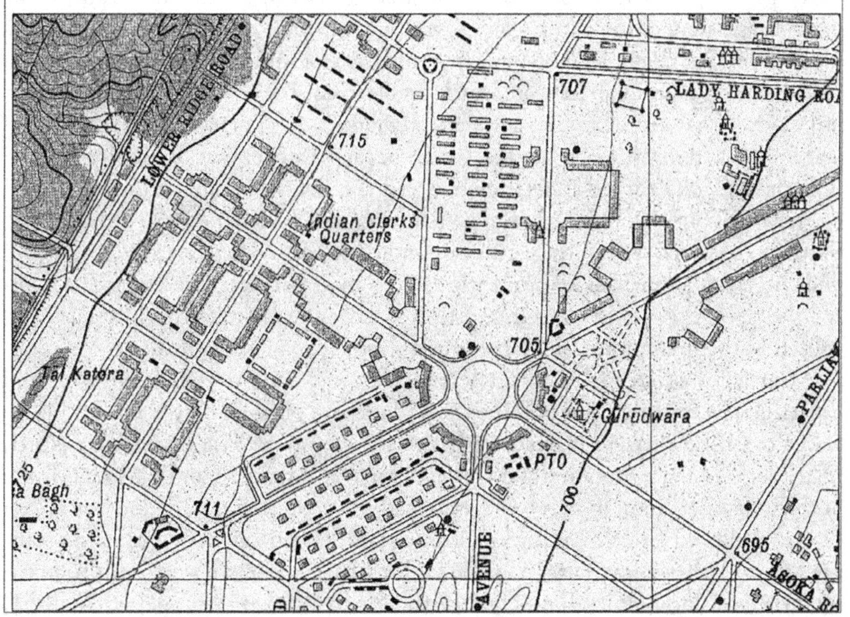

Figure 3.7 Built fabric of Shahjahanabad and New Delhi in 1922

Source: Calcutta: Survey of India Office (1922)

> suburb of broad streets and fine buildings known as New Delhi. Now that the Yanks have come to India, there has come into being a community, which may well be called the ninth Delhi.
>
> (Mehra, 2013, p. 360)

The floating populating was certain to influence the physical environment and the requirements of war forces. There was a sudden increase in housing demand, and Mehra (2013) explains that

> [w]ith the coming of the war, in the open spaces of Lutyens' Delhi which still awaited development, a new cheaply-built, modular and temporary architecture appeared in the form of tents, barracks and hutments for soldiers, officers and offices, as Indians, British, Americans, Australians, Canadians and others swarmed the city, bringing a "mushroom spawning of temporary huts for housing the ever-expanding Services headquarters."
>
> (p. 360)

Immigration of soldiers was also raising demand for daily-use items, and the opportunity was materialised by the local people, and many formal and informal production units were established in and around Old Delhi between 1939 and 1945 (Mehra, 2013). Inflow of foreign soldiers was a temporary phenomenon, but ever since then the immigration was continuously influencing the urban form of Delhi (refer to Chapter 4 for more details on immigration in Delhi). Ironically, these organic and temporary features are now the permanent characteristics of Delhi's urban form. For example, many informal industries continue to provide economic base to the city, although they are less desirable inside the city. The bazaar of Chandni Chowk has been a popular commercial market for retail and wholesale dealers and formal and informal businesses, which occupy every corner of the street.

In 1947, India gained independence and the British administration machinery left the country. The next wave of immigration was experienced due to the Partition of India, when thousands of Hindu and Sikh refugees arrived in Delhi, and there was a phenomenal increase of 90% in population (see Figure 3.8) during the decade 1941–51. It is interesting to note the changes that the Partition caused to the urban form of Delhi, which started expanding quickly.

After a lot of struggles to settle refugees, the Indian government started creating new resettlement colonies in and around New Delhi (Alluri and Bhatia, n.d.). The process started with the formation of Vijay Nagar in North Delhi as the first permanent settlement for the refugees (ibid). Model Town and Kingsway Camp, which was converted into 'Guru Teg Bahadur' (or GTB) Nagar, followed (ibid). Central Delhi was not left untouched, and many refugees moved into empty flats in Lodhi Colony (ibid). Houses were built in Nizamuddin and Jangpura villages. Also, the Khan Market was opened in 1951, where many shops were owned by refugees who started living in flats above the shops (ibid). The government bought agricultural land in South Delhi and created the residential colonies of Lajpat Nagar and Defence Colony and allocated land in Malviya Nagar (further southwest) for industrial developments (ibid). A similar approach was adopted to develop West Delhi, where

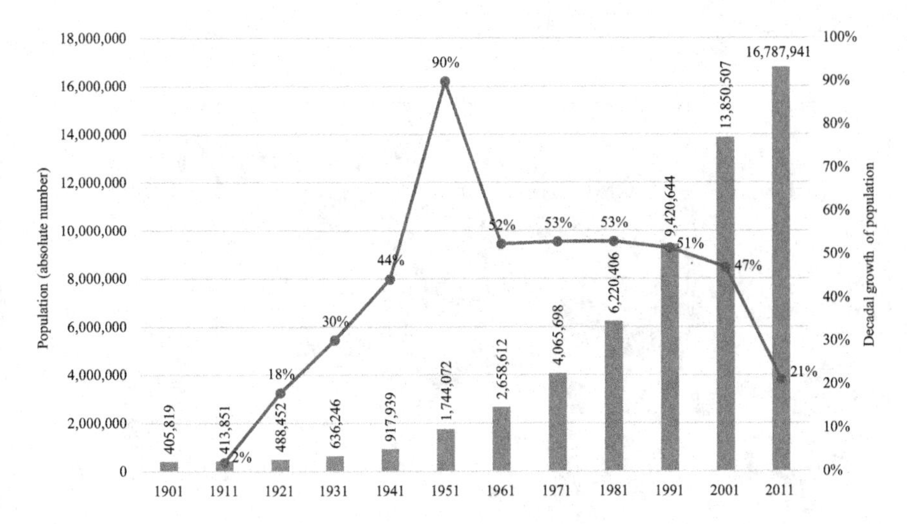

Figure 3.8 Decadal population of National Capital Territory of Delhi (1901–2011)

Source: Authors; data sourced from the Census of India (2011)

residential neighbourhoods were characterised by a U-shape layout with a park in the centre (ibid). These colonies were Rajinder Nagar, West Patel Nagar, Moti Nagar and Rajouri Gardens (Alluri and Bhatia, n.d.). By 1950, the government was able to house 300,000 refugees, of whom 190,000 were placed in evacuee houses (former Muslim properties) and 100,000 in the new colonies.

Despite government initiatives to provide housing to the immigrants, many refugees remained as 'squatters'[8] on government land which was not developed. They constructed temporary and permanent houses for themselves. The decade after Partition is summarised by Alluri and Bhatia (n.d.):

> A decade from independence, Delhi was a different city. Wilderness and agricultural fields began to give way to residential suburbs, commercial markets and industrial zones. The population doubled: a spurt that hasn't been seen since. . . . But the Muslim share of the population plunged from 33 percent to less than 6 percent.

Punjabi immigrants constituted major portion of Delhi's population. Alluri and Bhatia (n.d.) quote V. N. Dutta (Indian historian), who says (in the context of dominant culture) that "the city that was once a Mughal city, then a British city, had by the 1950s emphatically become a Punjabi city". Mehra (2013) finds that alongside planning and policy interventions, Delhi's urban growth was predominantly guided by "the conjuncture of global events, state actions, and the agency of citizens".

Along with Punjabis, there were Bengali immigrants coming from East Pakistan. A special area called 'East Pakistan Displaced Persons' (EPDP) colony was created in South Delhi in the early 1960s, which is now called the Chattaranjan Park, or CR Park. The area has developed well and is among the preferred locations for residence by the Bengali community.

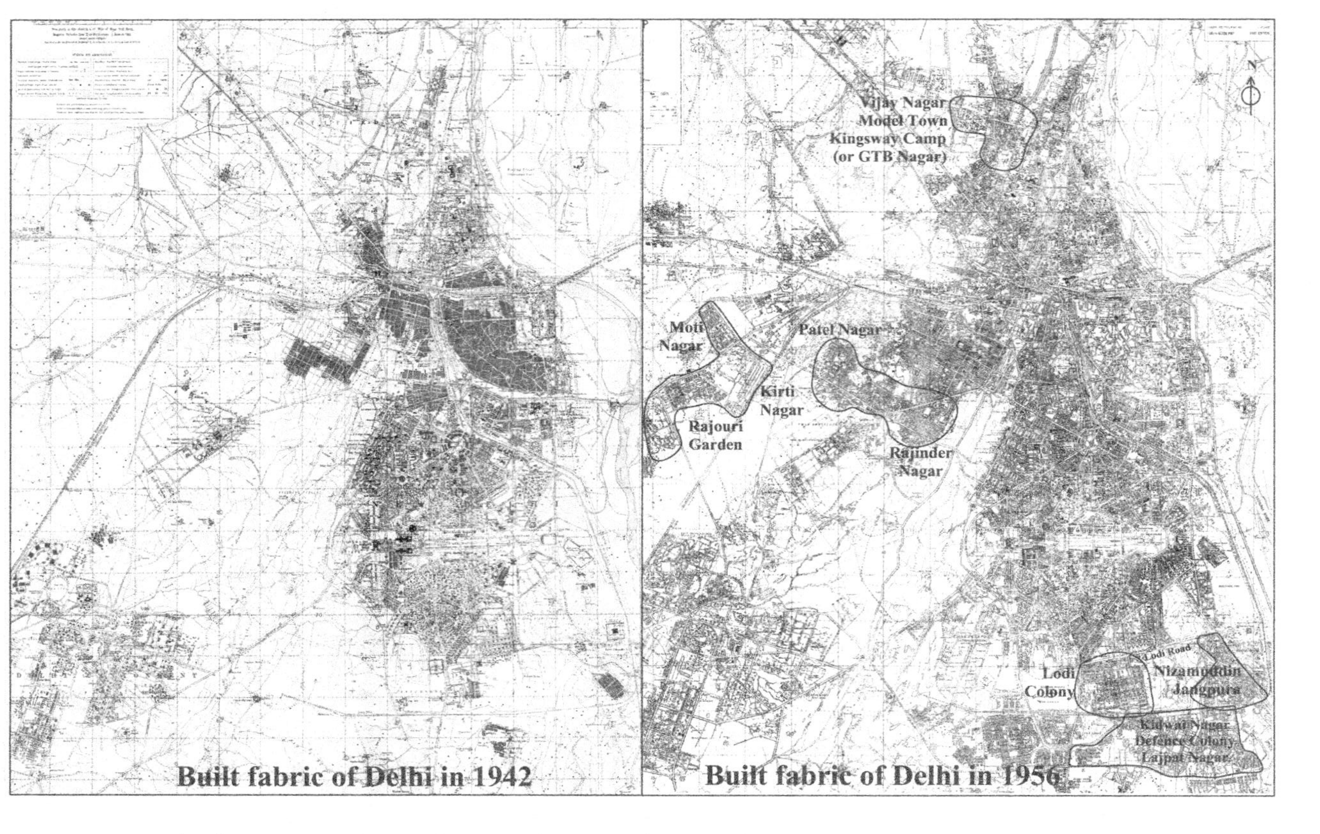

Figure 3.9 Map of Delhi in 1942 and 1956, showing refugee resettlement colonies post-Partition

Source: Authors adaptation from Alluri and Bhatia (n.d.)

Most of the city, as it exists now, is built upon these initial constructions which were undertaken to meet an urgent situation, thus allowing less time to plan. The aforementioned changes to Delhi's built environment mark the beginning of many squatter settlements.

3.10 Lal Dora

Another interesting feature that was born as an outcome of hasty development around Delhi was the 'Lal Dora' areas. These were the village settlement areas that were engulfed in urban boundaries as and when the city limits expanded. However, these pockets were treated differently and remained disintegrated from the physical developments happening around them (Seth, 2007). In the absence of a proper structure of transformation from rural to urban, many settlements have regressed to become urban slums[9] (ECLD, 2007). The following are the typical stages of degeneration:

Stage 1 – Village settlement area is surrounded by agricultural fields, which are also the primary source of income.

Stage 2 – The village is compulsorily acquired by the government, and urban boundaries are expanded. The planning authority demarcates an extended boundary for the village settlement area to grow in the future. Livelihood dependency on agricultural land is ignored, and there is no arrangement for alternate sources of income.

Stage 3A – Gradually, agricultural land around the settlement is developed while the character of the village is still retained with some farming and animal husbandry. At times, a ring of temporary residential structures develops into the zone of expansion. These are usually created by immigrants who otherwise do not have many opportunities to rent a house in the formal sector.

Stage 3B – The opportunity of nearness with urban society is used by the rural community who enter into small-scale commercial and industrial activities. The village periphery is thus a commercial ring, followed by the temporary settlers and the original settlers in the core. As the population inside these settlements is growing, these settlements are converting into low-rise, high-density slums.

The inclusion of rural land into urban boundaries is a usual phenomenon of urbanisation. There is an urgent need to derive a format in which village settlements can merge comfortably in urban boundaries and are given the time and facilities to adjust, economically and socially, into the new setup.

3.11 Master Plan of Delhi – 1962, 2001 and 2021

By 1955, the economic migrants from other states also started flowing into Delhi which was offering economic opportunities to the educated as well as the uneducated labour force (ibid). This is an ongoing trend. With intentions to provide a

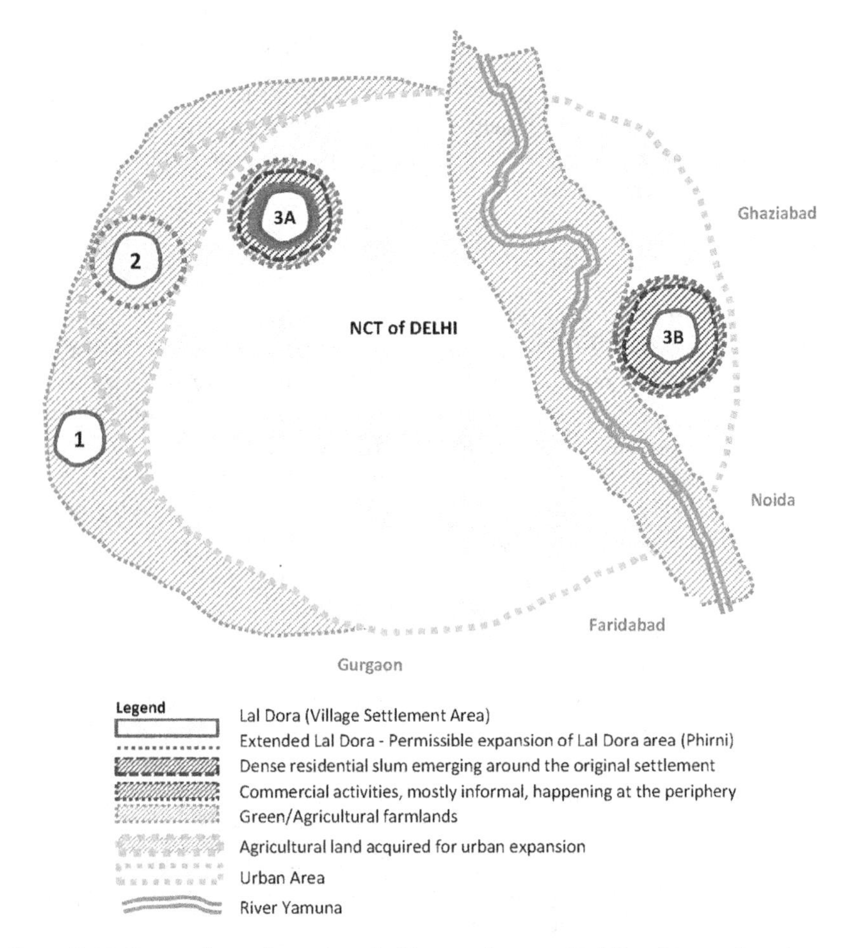

Figure 3.10 Stages of transformation of village settlements (Lal Dora) into urban slums

Source: Authors

Figure 3.11 Streets of Chilla village (Lal Dora area in East Delhi)

Source: Authors, 23 February 2014

planned and sustainable response to population growth the Delhi Development Authority was formed in 1957. The first Master Plan for Delhi was prepared in 1962 in collaboration with the Ford Foundation consultant team, which adopted a North American approach to planning (Priya, 2006).

The issues of congestion and overcrowding in Shahjahanabad (or Old Delhi) were acknowledged in the first development plan of 1962, but little success was achieved in relocating the non-conforming industries and in meeting the demand for commercial space (Delhi Development Authority, 2006). The growing demand for commercial space has been met by unauthorised mixed-use developments in residential areas, to the extent that front setbacks of the houses are covered and used for commercial purposes (ibid). Formal and informal activities in this area have been overruling planning proposals and were dominantly determining the physical pattern of growth of this area. In addition to that, planners overlooked the existing informal settlements in Delhi, which had grown exponentially, thus overloading infrastructure facilities (ibid). Even otherwise the population grew at a faster rate than estimated by planners, and this further caused unplanned densifications and infrastructure failures (Delhi Development Authority, 2006).

Today the social and spatial marginalisation of the poor is visible, as is the disintegration of Lal Dora, in Shahjahanabad and other similar areas that are identified with the urban poor. A major part of Delhi's population, around 30% to 40% of households, live in squatter settlements (Delhi Development Authority, 2006) and are mostly employed in informal sectors. However, there is wilful ignorance of the urban poor in the urban form of the city, probably because of the impatient urge to achieve the 'world-class' status for Delhi and thus the prioritisation of the aesthetic appearance, similar to other capital cities of the developed world which are "spacious, uncluttered, efficient, ordered, green, offer grand views – particularly of state and civic buildings – are clean, and do not contain poor people or informal activities" (Watson, 2009, pp. 174–175). While explaining the theory

Figure 3.12 View of Shahjahanabad (or Old Delhi) from Jama Masjid

Source: Authors, 14 January 2014

of urban modernism and its inappropriate application on poor and informal cities (like most parts of Delhi), Watson (2009) writes that

> the most obvious problem with urban modernism is that it fails to accommodate the way of life of the majority of inhabitants in rapidly growing, and largely poor and informal cities, and thus directly contributes to social and spatial marginalisation.
>
> (p. 175)

The first development plan was revised in 1990 and was upgraded to meet the requirements of the city till 2001. The new plan acknowledged mixed-use developments and selectively allowed it to happen in a few areas. Probably this was a reactionary approach to regularise existing mixed-use developments. The most interesting principle of this master plan was to decongest the city and create multiple nodes of developments around the city. This was to happen through the formation of 'National Capital Region' (NCR) that includes other cities from neighbouring states of Uttar Pradesh, Haryana and Rajasthan. Economic developments, together with the introduction of multimodal mass transport system (metro trains and ring roads) in Delhi, have resulted in 'floating population' between Delhi and other NCR cities. However, these cities are still not attractive enough to provide effective countermagnets for activities and people to Delhi, thus being inefficient in controlling migration to Delhi. On the contrary, satellite cities are off-shouldering their population into Delhi, thus causing higher migration than before (Delhi Development Authority, 2006).

Out of the total land area of 1,483 sq. km., Delhi had approximately 53% (or 798 sq. km.) rural land in 1991 (Government of NCT of Delhi, 2014). However, in the past two decades (1991–2011), the city has sprawled and engulfed rural land, thus increasing the urban land area by 30%, against an urban population increase of 8% (ibid). This is indicating the unsustainable spatial expansion of

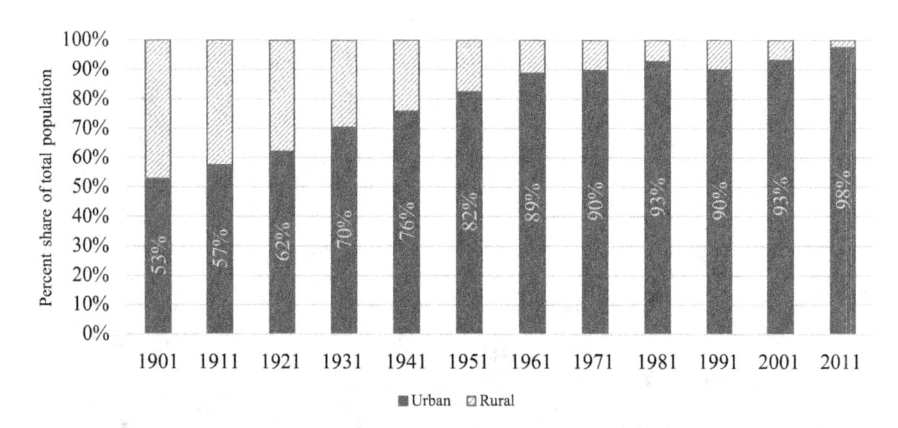

Figure 3.13 Urban and rural population as a share of the total population of Delhi, 1901–2011

Source: Authors; data sourced from Government of NCT of Delhi (2012)

the city that is caused by unruly conversation of agricultural land into mostly unauthorised developments in the urban periphery. Watson (2009) finds that there is ambiguity in the categorisation of certain developments as authorised and others as unauthorised and the rules are rather flexible for the wealthy and politically influential. Roy (2009) argues that "almost all of Delhi violates some planning or building law, such that much of the construction in the city can be viewed as 'unauthorised'" (p. 80). In addition to that, the change of use of peripheral agricultural land for industrial purposes or other real estate developments is influenced by private parties, more so because of the unclear land titles due to which formal transactions of land are constrained and less is available for development, thus causing speculations.

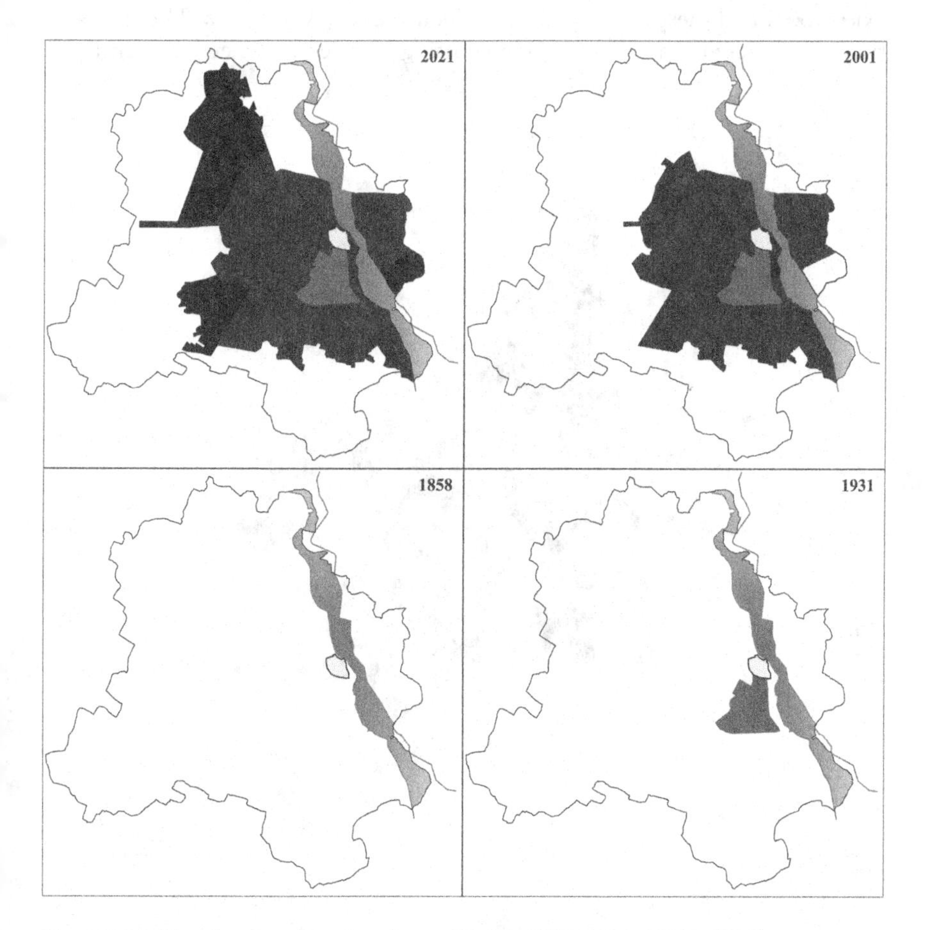

Figure 3.14 Transforming the urban form of Delhi, 1858, 1931, 2001[1], 2021[2]

1 As per the Master Plan of Delhi (MPD) 2001

2 As proposed under the Master Plan of Delhi 2021

Source: Authors

Widespread illegality and corruption in the government system and the irresponsible approach of planners are other reasons for the patchy urban form constituted by the planned and unplanned urban spaces (Mehra, 2013).

The latest Master Plan of 2021 acknowledges the aforementioned challenges and also identifies new determinants that may impact upon Delhi's built form, the prime ones being the increase in the aging population and the increase in workforce participation, particularly among females (Delhi Development Authority, 2006). It will be interesting to observe the performance of the new plan against these natural drivers.

These discussions touch upon important events in the history of Delhi that have caused an impact upon its built form. This information helps in identifying the non-design determinants of urban form. The objective here is to create a framework that allows interpretation of the urban form of Delhi as it exists today and these factors will continuously evolve with time. The next section will conclude the chapter by presenting a correspondence between these

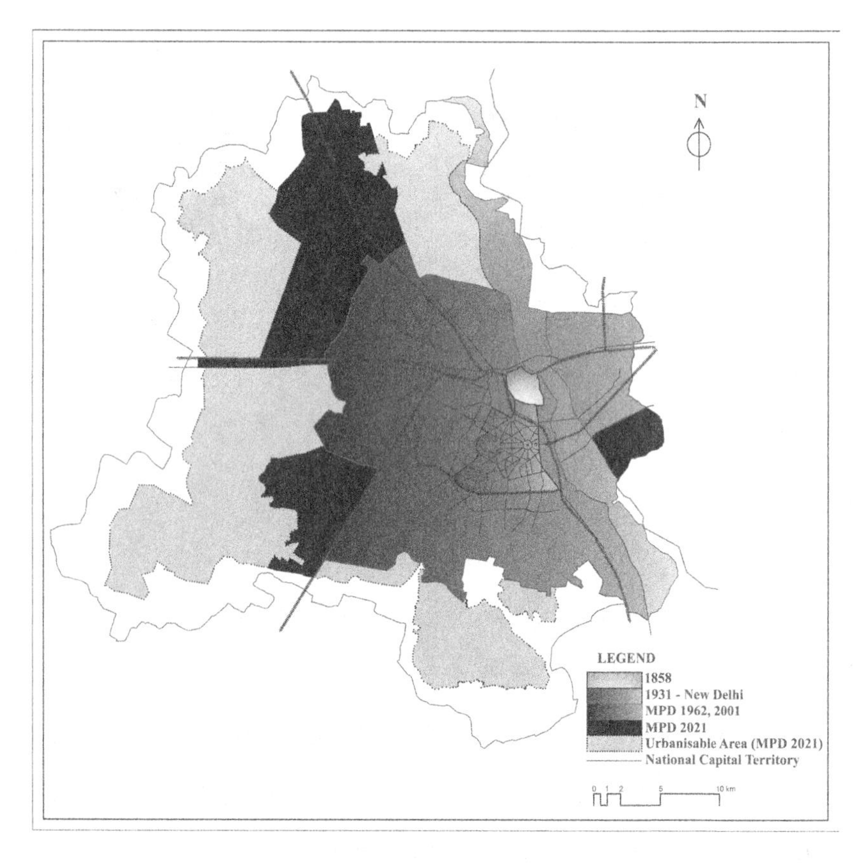

Figure 3.15 Urban form of Delhi, 1858–2021

Source: Authors

determinants and the original factors that were identified by Trigger (1968) for prehistoric cities.

3.12 Conclusion

"No culture has received so perfect an adjustment to its environment that it is static" (Lal, 1984). The processes that shape the city are complex, and its form is dynamic. Alongside design elements, there are multiple non-design factors which determine the urban form of a city, as discussed earlier in the case of Delhi. The focus of preceding discussions is around events and issues which have caused an influence on the urban form of Delhi. A combined list of modern and historic determinants is presented in Table 3.1.

The summary in Table 3.1 presents an interesting synthesis of history and modernity, exactly the way Delhi is. It is interesting to note that the form of Old Delhi has been guided by historic and modern determinants alike. A few historic determinants have evolved over time but are still important in the context of the modern city. One such example is the choice of location based on family and kinship ties. Although this may require further investigation, a tentative relationship can be drawn between family and community ties and people's choice of location. Many new determinants are also identified, and these together demand a broader discussion on urban theory of transforming city of Delhi.

Based on the discussions in this chapter, some important observations can be made in reference to Delhi's urban form:

> Delhi has a long political history and there are varying level of carry forward effects of it on different aspects of city's life, including its economy, society and the built form. Very careful analysis of each of these elements of city life is required for accurate estimation of the direction in which the city will grow and expand. This could guide designers and planners, whose role is to provide a structured spatial format in which these natural processes of city life can fit comfortably. This role should not be confused with that of fixing urban issues, "the root cause of which lie in broader institutional, political, socio-economic and environmental forces".
>
> (Watson, 2009, p. 189)

As mentioned in Chapter 1, each layer of the built environment (refer to Figure 1.2) has its own timeline of transformation. Urban form thus needs to acknowledge that the past has a momentum and the future form of the city should allow the transformation to happen. While it is good to set high aspirations and take inspiration from urban theories of the developed world, it is also necessary to draw the city in gradually transforming frames, each of which is not too far away from the immediate past.

Last but not the least, achievement of desired urban form is dependent upon satisfactory operation of polity, economy and society, in the absence of which its realisation is diminished.

Table 3.1 List of modern and historic determinants of urban form of Delhi

Determinants of settlement pattern of prehistoric city	*Factors influencing the urban form of modern Delhi*	*Explanatory remarks*
	At the city level	
Size and location of a city are determined by -		
Availability of natural resources for survival Availability of natural resources for economy		From arguments presented by Spear (2002), it may be concluded that the ecology of Delhi was not the most attractive feature. However, there is no information available about the settlements in the city prior to Rajput, who had political reasons to locate at Delhi.
Family and kinship organisations	This is furthered by regional bonds and community ties	Delhi is spatially fragmented into many communities. For example – Old Delhi is still holding its native population which is primarily Muslim. Refugee colonies can be easily identified with Punjabis and Bengalis. Lal Dora areas are the village communities. These communities are strongly bonded together and prefer to be nearly located.
	Political control and efficient administration	Delhi was chosen as Mughal capital since it was central to Afghanistan and India, thus making it easy for Mughal rulers who were ruling over both the regions.
	Weather and climatic conditions	For its pleasant weather, Shimla was chosen as the summer capital for the British.
	Economic opportunities	Calcutta, being a port city, was the base of trade and commerce for the East India, Company and was the British capital city for many years. Transport connectivity between Delhi and the NCR has introduced a floating population from Delhi to the surrounding cities which are offering work and employment.
	Quality of infrastructure and governance	Immigration in Delhi is increasing, particularly after its connectivity with other NCR cities. People are choosing to live at Delhi, which offers relatively better infrastructure and governance than many other cities in North India.

Determinants of settlement pattern of prehistoric city	*Factors influencing the urban form of modern Delhi*	*Explanatory remarks*
The layout of the city is determined by -		
Family and kinship organisations	This is furthered by regional bonds and community ties Spatial segmentation of different income groups	However, its influence on the layout of the city is immaterial in modern city planning.
Public buildings in the centre of the city		The cities of Shahjahanabad and New Delhi were laid out, placing major administrative buildings in the centre.
Level of complexity of political organisations and defence system		The city of Shahjahanabad was contained within its walls, and this made it easy for the ruler to safeguard its subjects. The consequences are that majority population inside walled city belongs to Muslim community, and the area inside walls is very densely populated with spillover effects in its neighbourhoods.
	Efficiency of governance and security for citizens	Private developers are continuously building large, gated communities on urban fringes of Delhi and NCR. These are the new "walled cities", where people of a certain income class usually come together for various reasons, the most important being physical security and good-quality infrastructure.
Stability of the government, political tensions and warfare		Continuous tensions between various kingdoms, foreign invasions and inter-dynasty tensions led to the creation of many different cities in Delhi, spanning over ancient, medieval and modern times. Later, the British relocated the capital from Kolkata to Delhi to avoid political tensions with the native Indians.
Cosmological beliefs		It is hard to comment upon the influence of cosmological beliefs on city layouts. Nevertheless, the concepts of vastu shastra are applied by many Indians, particularly in residential layouts.

(*Continued*)

Table 3.1 (Continued)

Determinants of settlement pattern of prehistoric city	*Factors influencing the urban form of modern Delhi*	*Explanatory remarks*
	City as a symbol of government ideologies, political strength, and racial superiority	Shahjahanabad and New Delhi are laid out keeping many such symbols, as explained in the chapter.
	Infrastructure corridors attract new developments alongside	New transport corridors between Delhi and NCR cities are experiencing unauthorised haphazard developments alongside them.
	Existing pattern of use	Since Shahjahanabad was used as the commercial centre for Delhi, there is continuous growth of commercial activities, thus an increasing demand for space. This includes many informal businesses by the local people, which has been the original character of this area.
	Ease of acquiring land	Challenges of acquiring land for the resettlement of refugees have created a patchwork of refugee colonies all over Delhi.
	Land tenure	Lack of clear land titles is restricting availability of land in the open market, thus causing superficial scarcity and speculations. The two flaws together, mixed with corruption, motivate private developers to acquire and develop agricultural land at urban fringes. This is leading to unregulated sprawl of the city.
	Approach towards village settlements which are included in urban boundaries	Village settlements (or Lal Doras) were included in urban boundaries without actually integrating them into the urban system. Unplanned haphazard growth of urban villages has changed villages' character to slums.
	At the regional level	
Ecology		So far Delhi has not run out of basic resources, and it may still have the capacity to support its growing population, provided environmentally sustainable practices are adopted.

Determinants of settlement pattern of prehistoric city	*Factors influencing the urban form of modern Delhi*	*Explanatory remarks*
Demography		Demographic changes in the long history of the city have caused a serious impact on its urban form. Immigration and population increases may continue to remain important determinants of its urban form in the future as well.
Political structure		Delhi was planned to be an independent federal city which should not be linked to any particular state or region. While every state has its own capital city, the supremacy of Delhi as the national capital is also visible in its urban form.
	Political ties with neighbouring states	Governance and administration might influence the settlement pattern to a good extent. As seen in the case of Delhi, its connection with NCR cities has not been very successful due to differential approaches adopted by governments.
Religion		The influence of religion on urban form is not significant. Many religious institutions own land, and also at times the government allocates land for the establishment of religious buildings.
Economy		Economic activities in the Delhi NCR region are together offering many opportunities for employment in public and private sectors and formal and informal sectors across all trades like service, IT, manufacturing, hospitality and retail. This is a strong pull factor, which is enhanced by the push factor of neighbouring states which lack seriously in economic development. In addition to that, there are many prestigious educational institutions at Delhi which together attract a lot of the students inside the city, thus creating employable labour force.

(*Continued*)

Table 3.1 (Continued)

Determinants of settlement pattern of prehistoric city	*Factors influencing the urban form of modern Delhi*	*Explanatory remarks*
	Characteristics of neighbouring regions	As discussed earlier, the neighbouring states to Delhi suffer from multiple issues of poor infrastructure, weak economy, political instability, social discriminations and the like. Delhi is a relatively better place for work and living, and therefore it attracts immigrants from the neighbouring states. With intentions to decongest the core areas of Delhi, other neighbouring cities are connected with Delhi, although it is creating a reverse osmosis and population is flowing into Delhi from the NCR region.
	Urban planning approach to population increase and future expansion	Certainly Delhi's approach to create satellite towns did not give the desired outcomes. One of the prerequisites for efficient application of this approach will be equal standards of living in all satellite towns and primate city. Failure of this approach in Delhi is another example that urban theories of Western cities may not perfectly apply in the Indian case.

Source: Authors

Notes

1 For the Census of India 2011, the definition of urban area is as follows:

1. All places with a municipality, corporation, cantonment board or notified town area committee, etc.
2. All other places which satisfied the following criteria: i) A minimum population of 5,000; ii) At least 75 per cent of the male main working population engaged in non-agricultural pursuits; and iii) A density of population of at least 400 persons per sq. km. The first category of urban units is known as Statutory Towns. These towns are notified under law by the concerned State/UT Government and have local bodies like municipal corporations, municipalities, municipal committees, etc., irrespective of their demographic characteristics as reckoned on 31st December 2009. Examples: Vadodara (M Corp.), Shimla (M Corp.) etc.

The second category of Towns (as in item 2 above) is known as Census Town. These were identified on the basis of Census 2001 data.

Source: (Census of India, 2011)

2 Basham (1969).
3 "uniformity of weights and measures, the common script, the uniformity – almost common currency – of the seals, the evidence of extensive trade in almost every class of commodity throughout the whole Harappan culture zone, the common elements in architecture and town-planning, the common elements of art and religion" (Allchin & Allchin, 1968, p. 129).
4 Kenneth Gillion has written in length on the city of Ahmedabad in the book titled *Ahmedabad. A Study in Indian Urban History*.
5 Good, Kenoyer and Meadow (2009).
6 "The Mesolithic, or in India the Late Stone Age, is the terminal hunting gathering stage (of human civilization). It follows the Palaeolithic or Old Stone Age. The Neolithic is the stage of early village life or domestication of plants and animals. When prehistoric farmers begin to use metal (copper) in a small way the stage is called the Chalcolithic, although there is little change in basic social or economic structures between the Neolithic and Chalcolithic stages. When, however, we use the term bronze age we generally mean a stage when there are specialist producers of metals and other products in society, that is, the earliest stage of social complexity. Thus where South Asia is concerned, the bronze age is the age of the Harappan Civilization" (Ratnagar, 2002, p. 3).
7 Another guideline prepared by Kautilya during the third century BC.
8 Squatter settlements are defined as illegal developments on public or private land without legal claims to the land.
9 This books adopts census definition of slum, which identifies slums "as residential areas where dwellings are unfit for human habitation by reasons of dilapidation, overcrowding, faulty arrangements and design of such buildings, narrowness or faulty arrangement of street, lack of ventilation, light, or sanitation facilities or any combination of these factors which are detrimental to the safety and health".

Bibliography

Acharya, P. K. (1980). *Architecture of Manasara: Illustrations of Architectural and Sculptural Objects* (2nd ed., Vol. V). New Delhi: Oriental Books Reprint Corporation.

Alexander, C. (1965). A City Is Not a Tree. *Architectural Forum, 122*(1,2), 58–62. Retrieved May 27, 2017, from http://webcache.googleusercontent.com/search?q=cache:hyjBiwlx6O8J:en.bp.ntu.edu.tw/wp-content/uploads/2011/12/06-Alexander-A-city-is-not-a-tree.pdf+&cd=1&hl=en&ct=clnk&gl=au

Allchin, B., & Allchin, R. (1968). *The Birth of Indian Civilisation*. Great Britain: Richard Clay (The Chaucer Press) Ltd.

Alluri, A., & Bhatia, G. (n.d.). *The Decade That Changed Delhi*. Retrieved May 28, 2017, from Hindustan Times: www.hindustantimes.com/static/partition/delhi/

Basham, A. L. (1969). *The Wonder That Was India: A Survey of the History and Culture of the Indian Sub-continent Before the Coming of the Muslims*. London: Sidgwick & Jackson.

Beglar, J. D. (1874). *Report for the Year 1871–72 (Delhi)*. Calcutta: Archaeological Survey of India

Blake, S. P. (1986). Cityscape of an Imperial Capital: Shahjahanabad in 1739. In R. E. Frykenberg (Ed.), *Delhi Through the Ages: Essays in Urban History, Culture and Society* (pp. 143–151). New Delhi: Oxford University Press.

Blake, S. P. (1991). *Shahjahanabad: The Soverign City in Mughal India, 1639–1739*. Cambridge: Cambridge University Press.

Calcutta: Survey of India Office. (1922). *Late 19th- and Early 20th-Century Asian Cities*. Retrieved June 15, 2016, from The University of Chicago Library: www.lib.uchicago.edu/e/collections/maps/asian-cities/

Census of India. (2011). Census of India 2011: Provisional Population Totals (Urban Agglomerations and Cities). Retrieved May 27, 2017, from Office of the Registrar General & Census Commissioner, India: http://censusindia.gov.in/2011-prov-results/paper2/data_files/India2/1.%20Data%20Highlight.pdf

Chang, K. C. (1962). A Typology of Settlement and Community Patterns in Some Circumpolar Societies. *Arctic Anthropology*, *1*, 28–41.

Childe, V. G. (1950, April). The Urban Revolution. *The Town Planning Review*, *21*(1), 3–17.

Cohen, R. J. (1989). An Early Attestation of the Toponym Dhilli. *Journal of the American Oriental Society*, *109*(4), 513–519.

Delhi Development Authority. (2006). *Master Plans*. Retrieved May 30, 2017, from Delhi Development Authority: https://dda.org.in/planning/master_plans.htm

ECLD. (2007). *Report of the Expert Committee on Lal Dora & Extended Lal Dora in Delhi*. New Delhi: Ministry of Urban Development.

Erdosy, G. (1985). The Origin of Cities in the Ganges Valley. *Journal of Economic and Social History of the Orient*, *28*(1), 81–109.

Frazer, J. (1995). *An Evolutionary Architecture*. London: Architectural Association.

Good, I. L., Kenoyer, J. M. & Meadow, R. H. (2009). New Evidence for Early Silk in the Indus Civilisation. *Archaeometry*, *51*(3), 457–466.

Government of NCT of Delhi. (2014). *Statistical Abstract of Delhi 2014*. New Delhi: Directorate of Economics & Statistics, Government of NCT of Delhi.

Hearn, G. R. (1997). *The Seven Cities of Delhi*. New Delhi: S.B.W. Publishers.

Imperial Gazetteer of India. (1909). *Imperial Gazetteer of India, v. 26, Atlas 1909 Edition, Delhi* (p. 55). Retrieved May 28, 2017, from Digital South Asia Library: http://dsal.uchicago.edu/reference/gaz_atlas_1909/pager.html?object=61

Johnson, D. A. (2015). *New Delhi: The Last Imperial City*. New York, NY: Palgrave Macmillan.

Khosla, R. & Rai, N. (2005a). Glory of Empire: Imperial Delhi. In R. Khosla (Ed.), *The Idea of Delhi* (pp. 44–53). Mumbai: Marg Publications.

Khosla, R. & Rai, N. (2005b). The City as an Idea. In R. Khosla (Ed.), *The Idea of Delhi* (pp. 8–21). Mumbai: Marg Publications.

Kishore, R. (2015). Urban 'Failures': Municipal Governance, Planning and Power in Colonial Delhi, 1863. *The Indian Economics and Social History Review*, *52*(4), 439–461.

Kropf, K. (1996). Urban Tissue and the Character of Towns. *Urban Design International*, *1*(3), 247–263.

Lal, M. (1984). *Settlement History and Rise of Civilisation in Ganga-Yamuna Doab, from 1500 B.C. to 300 A.D.* New Delhi: B.R. Publishing Corporation.

Marshall, S., & Caliskan, O. (2011). A Joint Framework for Urban Morphology and Design. *Built Environment*, *37*(4), 409–426.

Mehra, D. (2013). Planning Delhi ca. 1936–1959. *Journal of South Asian Studies*, 36(3), 354–374.

Murdock, G. P. (1949). *Social Structure*. New York, NY: The Macmillan Company.

Oliveira, V. (2016). *Urban Morphology: An Introduction to the Study of the Physical Form of Cities*. Switzerland: Springer.

Priya, R. (2006). Town Planning, Public Health and Delhi's Urban Poor: A Historical View. In S. Patel & K. Deb (Eds.), *Urban Studies* (pp. 223–245). New Delhi: Oxford University Press.

Rakesh. (2014). Delhi Under the British Rule from 1803 upto 1911. *International Research Journal of Commerce Arts and Science*, 5(1), 316–331.

Ratnagar, S. (2002). Archaeological Perspectives on Early Indian Societies. In R. Thapar (Ed.), *Recent Perspectives of Early Indian History* (pp. 1–59). Mumbai: Popular Prakashan Pvt. Ltd.

Ray, A. (1964). *Villages, Towns and Secular Buildings in Ancient India: c. 150 B.C.–c. 350 A.D.* Calcutta: K.L. Mukhopadhyay.

Redfield, R., & Singer, M. B. (1954). The Cultural Role of Cities. The Role of Cities in Economic Development and Cultural Change, 1, 53–73.

Roy, A. (2009). Why India Cannot Plan Its Cities: Informality, Insurgence and the Idom or Urbanisation. *Planning Theory*, 8(1), 76–87.

Seth, S. (2007, February 4). Historical Transformations in Boundary and Land Use in New Delhi's Urban Villages. *Economic and Political Weekly*, *LII*(5).

Sinopoli, C. M. (1994). Monumentality and Mobility in Mughal Capitals. *Asian Perspectives*, 33(2), 293–308.

Spodek, H. (1980). Studying the History of Urbanisation in India. *Journal of Urban History*, 6(3), 251–295.

Thapar, R. (1987). *Ancient Indian Social History: Some Interpretations*. New Delhi: Orient Longman Limited.

Thapar, R. (2002). The First Millennium B.C. in Northern India (Up to the End of the Mauryan Period). In R. Thapar (Ed.), *Recent Perspectives of Early Indian History*. Mumbai: Popular Prakashan.

Trigger, B. G. (1968). The Determinants of Settlement Patterns. In K. C. Chang (Ed.), *Settlement Archaeology* (pp. 53–78). Palo Alto, CA: National Press Books.

Trigger, B. G. (1978). *Time and Traditions: Essays in Archaeological Interpretation*. Edinburgh: Edinburgh University Press.

Watson, V. (2009). The Planned City Sweeps the Poor Away: Urban Planning and 21st Century Urbanisation. *Progress in Planning*, 72, 151–193.

Weller, E. (1858). *1863 Dispatch Atlas Map of Delhi, India*. Retrieved May 28, 2017, from Geographicus Rare Antique Maps: www.geographicus.com/P/AntiqueMap/delhi-dispatch-1867

4 Continuities and discontinuities

4.1 Introduction

Having discussed the changing urban form of Delhi and its determinants in Chapter 3, this chapter uncovers the next layer of urban tissue of Delhi, which is the buildings and their elements. This chapter explores the various stages of transformation of buildings and architectural styles, from ancient to modern, including discussions on the commonalities that were carried forward and the new elements that were introduced at different points of time. There is ample information available on architectural details of almost all important historic and modern buildings that exist in and around Delhi. The chapter also knits together the building with its corresponding social, political and natural environment and identifies non-design determinants of design, to the extent feasible with the secondary sources of information. This may sound ambitious at this stage when there is a paucity of work that jointly explains the process of formation of buildings together with the design. This chapter initiates the discussions on amalgamating the study of process alongside that of the product (or the building) with the hope that this would result in a comprehensive understanding of Delhi's built environment.

Smith (2012) writes that before Second Industrial Revolution of the late nineteenth and early twentieth centuries, ornamentation was central to architecture to the extent that the study of architecture was in fact the study of ornament. Discussions on architectural styles were basically centred on different ornamentation styles – Moorish, Romanesque, Roman, Greek and similar others (ibid). Therefore, most literature on architectural styles of buildings before the nineteenth century is focused on ornamentation. This is valid in the case of Indian architecture as well, and the majority text discusses the ornamentation of Hindu, Buddhist, Jain and Islamic architecture. There is little inquiry about the reasons for the similarities and dissimilarities in various styles of architecture that were adopted during different time periods and which now coexist as a living history of modern Indian cities. While acknowledging the limitation of inadequacy of information available on the determinants of architectural styles (including social, political, economic, religious, cosmological or any other), this chapter widens the discussion to include the reasons, to the extent possible, for such continuities and discontinuities in the architectural styles as observed in the city of Delhi. In the dearth of non-literary evidences, any discussion on this topic is challengeable and merely probabilistic. The present study is also prone to such criticisms though with the following logic presented we would try to minimise that.

4.2 Pre-Islamic architecture at Delhi

There are few physical evidences remaining of ancient Delhi, which limits our understanding of the architectural style prior to Muslim invasion. Ancient Hindu evidences in Delhi, as listed by Beglar, are Fort of Anangpal or the Lal Kot; the iron pillar from fourth century which was probably brought to Delhi and installed at Lal Kot by Anangpal; the structure now called the Quwwat-ul Islam Mosque; and the structure called Sultan Ghari Tomb (Beglar, Carlleyle & Cunningham, 1874).

The fort of Lal Kot is the oldest remains of pre-Muslim Delhi which was built around 1052 AD by Rajput Tomar ruler Anangpal (Sharma, 2001). With some reservations, Beglar, Carlleyle and Cunningham (1874) conclude that Anangpal's Lal Kot was a small defence citadel (Figure 3.2) that did not enclose the city on the eastern, lower side of the citadel. The wall of the fort was constructed of large undressed stones put in rubble masonry using red earth mortar. The walls are surrounded by ditches of varying widths, 18 to 35 feet. Houses were made on brick foundations, and no superstructure remains to explain more. Peck (2005) explains that important buildings such as those built for defence and court purposes or religious purposes would have been built in superior and durable material as compared to domestic structures. The walls of Lal Kot were later expanded by Prithviraj Chauhan who constructed Qila Rai Pithora and enclosed the entire city inside fort walls. However, there is a paucity of information on precise design and layout of fort and settlements around it, and therefore its purpose, use and building characteristics remain unknown (Beglar, Carlleyle, & Cunningham, 1874).

Acknowledging the challenge posed by the limited information available on physical structures in Delhi before the arrival of Islam, the discussion on Delhi's built environment and architectural style will start from Islamic constructions. Occasional reference to Hindu architecture will be made in the course of discussion.

4.3 Islamic architecture at Delhi

The Islamic architectural history in India is often studied under three distinct phases differentiated by the changing political regime: first, the Delhi Sultanate (1206–1526 A.D.); second, the provincial period; and, third, the Mughal period (1526–1707 A.D.) (Sharma, 2001; Luniya, 1978; Khan, 2016). Differences in the political environment during these periods, explained in Chapter 1, had a significant influence on the architectural style, as will be discussed in this chapter. In summary, the first phase or Delhi Sultanate phase marks the beginning of Islamic rule in India with the establishment of Slave dynasty between early thirteenth (c. 1206) and sixteenth centuries (c. 1526). However, this was a period of political turbulence and warfare, and very limited construction activity happened during this time (Sharma, 2001). Internal conflicts resulted in frequent change of dynasties, each having its own individual architectural style (ibid). Political expansion was also limited to Delhi and Agra, and therefore most of the buildings of this period are located in this region (Sharma, 2001). In the second phase, which overlaps with the first and the third, many provincial architectural styles were founded by Muslim governors who declined allegiance with Delhi Sultanate and established independent provincial empires (ibid). The architecture in these provinces is in accordance with the individualistic likings of the provincial rulers (ibid). Later, in the third phase starting in sixteenth century, the Mughal rule was founded by Babur, and a vast empire was established in India. This was a peaceful and prosperous time, and significant architectural contribution was made, with respect to both quality and quantity (ibid). Delhi, being the imperial capital city for a long time during the Mughal period, enjoyed substantial architectural patronage. While provincial architecture has no influence on Delhi, the monuments mostly belong to either the first or third phase and discussions in this chapter will be limited to these.

4.4 Continuities and discontinuities – Hindu and Islamic architecture

With the advent of Islam in India during the Muslim rule from twelfth century AD the native Indian population was interacting with a foreign entity for the first time and the scope of negotiations, in the political, social, religious and architectural contexts, was yet to be explored. Architectural expressions of Hindus and Muslims (referred as Islamic architecture) were driven by many non-design factors including the following: (i) spatial requirements of the religion, the polity, and the society; (ii) religious and cosmological tenets; (iii) admissibility and popularity of various styles of ornamentation; (iv) inspiration and mimicry of building techniques and ornamentation from the past; and (v) technical knowledge of structural systems and ability to use different building materials. While this section introduces these factors broadly, the next sections will keep referring back to these factors and will provide more explanations as and when the discussions advance.

4.4.1 *Spatial requirements*

The requirement of space for religious, political and social purposes are often served by dedicated building types. An interesting finding by Ray (1964), in the context of ancient societies, suggests that the diversity of functional typology of buildings is synonymous to the level of complexity of the society (ibid). This can probably be expanded in the social and religious context to mean that more fragmented the society is, the higher is the exclusivity of use of buildings, thus increasing the diversity of functional typology in buildings (ibid). For example, the differences in schools of theologies of Islam gave rise to the establishment of *madrasa* as an Islamic educational institution, which often was used to propagate the theological ideology of the patrons (Tabbaa, 2001).

Prior to the advent of Muslims in India, the built environment of the country was primarily constituted by temples, stupas, monasteries, shrines, palaces and forts (Burton-Page, 2008) and also stepwells and waterbodies. But almost none in Delhi are worth mentioning from architectural perspective. Often, Hindu architecture is assumed to be synonymous with temple architecture and other religious monuments, and the emphasis on secular buildings is very limited. This is not to be inferred as absenteeism of non-temple building constructions in the native Hindu society and instead is cautiously interpreted by Ray (1964) to mean that temples, being the abode of immortal God, were probably constructed of durable materials which outlived time and survived multiple waves of invasions, while other structures, made from perishable materials, are extinct.

With Islam were introduced new building types in India and in Delhi in particular. The new needs of this foreign society demanded various types of religious and secular buildings, including mosques, tombs and cenotaphs in the first category and miscellaneous public and civic buildings in the secular category, like palaces, forts, gardens, town gates, wells and reservoirs, pavilions and even entire cities (Brown, 1956).

4.4.2 *Religion and places of worship*

Brown (1956) writes that "nothing could illustrate more graphically the religious and racial diversity, or emphasise more decisively the principles underlying the consciousness of each community than the contrast between their respective places of worship" (p. I). The foundation of the difference in the layout of temple and mosque lies in the differences in the religious beliefs and thus differential requirements for spaces, as discussed next.

A temple is symbolised as the temporary abode of God on earth where a devotee seeks the opportunity to communicate with the lord. Prayer to the God is a form of personal communication which happens behind closed doors in a private room in which the God resides, called the *garbha griha* or the womb. The privacy of this room is guaranteed with limited openings, which restricts sound and visibility. Thus, *garbha griha* is usually a small, dark cell which is 'jealously guarded' (Brown, 1956, p. I). Being the abode of the lord, the interiors and exteriors of Hindu temples are covered in figurative decorations which usually showcase religious stories associated with the God of the temple, as mentioned in the religious corpus. Also, it is customary for a devotee to face in the east direction while praying and performing religious ceremonies, and thus most temples are aligned east-west with *garbha griha* towards the east.

The basic design of a mosque takes inspiration from the house of the Prophet Muhammad in modern-day Saudi Arabia (Grover, 1996) (Figures 4.1 and 4.2). The courtyard of the house served as the gathering place for the faithful (ibid). The tradition of congregational prayer is carried forward, and the courtyard (or *shin*) forms an essential component of mosque design. In addition to that, a semi-covered colonnaded prayer area (or *riwaq*) is provided on one or all sides of the courtyard, like in the house of the Prophet Muhammad (Othmann, Aird & Buys, 2015). While offering prayer, the worshippers face towards Mecca, and the western wall (or *qibla*) identifies the direction (Sharma, 2001). This is often differentiated from other walls by an ornamented recess in the centre (or *mihrab*) (ibid). A covered prayer hall (or *liwan*) is also attached to the western wall (ibid). The roof of the hall is domical, probably as a symbol of the vault of heaven (ibid). This notion is reinforced due to iconographic ornamentation of the dome, as will be discussed later. As a structural element the use of dome in the prayer hall reduces the number of columns, thus providing the wider column free space for gathering (Gye, 1988). The use of an arcuate system in general allowed for the creation of bigger spaces as compared to the trabeated systems of Hindu temples (ibid).

Therefore, the fundamental difference in the design of temple and mosque lies in the religious process of personal prayer of Hindus as opposed to congregational prayer of Muslims.

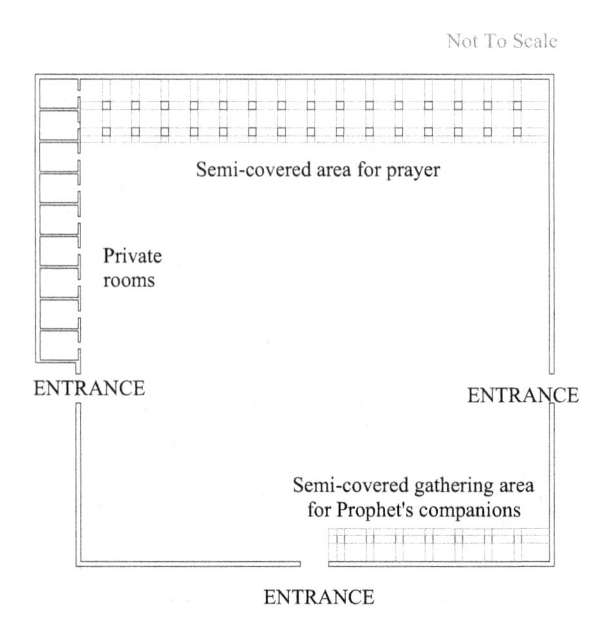

Figure 4.1 Sketch plan of the Prophet Muhammad's house in Medina (623 A.D)

Source: Recreated by authors with base image from Othmann, Aird, and Buys (2015)

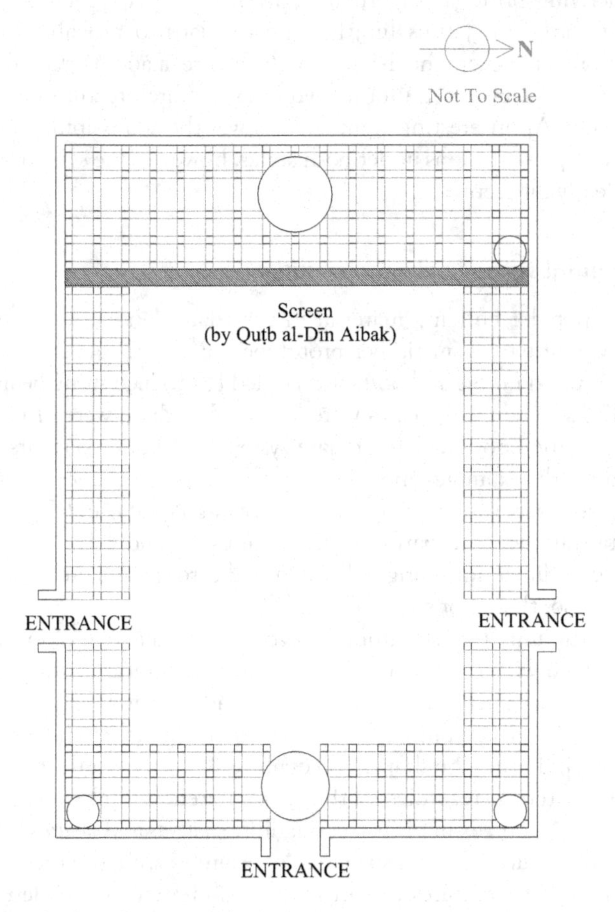

Figure 4.2 Sketch plan of Quwwat-ul Islam Mosque

Source: Recreated by authors with base image from Murray (c. 1911)

4.4.3 Theology and ornamentation

Tawhid is the defining principle of Islam which declares absolute monotheism or the oneness of God (*The Oxford Dictionary of Islam*). One of the interpretations of Islamic monotheism, adhered by Mutazils (a school of Islamic theology), is that God is divested of all human attributes because anthropomorphism constitutes a form of plurality ('shirk'), thus opposing the very essence of Islam (Tabbaa, 2001, p. 12). This principle mandated non-figurative ornamentation as an essential characteristic of Islamic architecture, particularly in the context of religious monuments (ibid). The restricted use of human and animal figures in palatial ornamentation is observed in Fatimid Cairo (ibid).

Without involving much into the religious discourses of Hinduism, it is easy to conclude that art and architecture of Hindus are dominated by figurative representations, and the identification of the god with a *murti* or idol is common.

The underlying principle of ornamentation of the two religion groups was significantly contrasting, thus limiting the overlaps to probably only vegetal ornamentation. However, the Islamic architecture made a gradual shift from naturalistic to abstract vegetal forms, while Hindu architecture was complying with the former. An interesting commonality was the admissibility for the use of iconographic representations under both styles; however, the symbols and meanings were clearly different.

4.4.4 *Structural system and ornamentation*

Stone was a popular building material in pre-Islamic India, and the structural technique was derived from timber prototypes (Burton-Page, 2008). Trabeated framework of wood rafters and posts was copied to produce stone beams and columns. Arches and domical shapes were known, but these were often solid constructions, and the knowledge of arcuate system and stone voussoirs was absent (ibid). Some early examples include the arch-shaped entrance door of Lomas Rishi Cave and barrel shaped or curved roofs at Ajanta and Ellora Caves in Maharashtra state, western part of India (Figures 4.3 and 4.4). These are imitations of the curvilinear form originally obtained through the use of pliable materials like bamboo, thatch or straw (Ray, 1964).

Similar to the imitative structural design, the Hindu style of ornamentation may be explained with the theory of mimetic ornamentation, which is the commonest type of architectural ornamentation among primitive societies in Asia (Ackerman, Gowans, & Collins, 2017). It is explained as the tendency of the artist to continue imitating the shapes and qualities known from the past and install these on new materials, regardless of the appropriateness (Ackerman, Gowans, & Collins, 2017). A good example is the Buddhist *chaityas* and *viharas* from the second and third centuries BC, for example, at Ajanta Caves (Figures 4.3 and 4.4) where the rock-cut structures are an accurate imitation of wooden prototypes known from the past (Ray, 1964). In many Buddhist stupas, like at Sanchi, the joineries and shapes of the railings are a direct translation of wooden details in stone, as seen in the tenons of the uprights and the scarf jointing of the copings, neither of which are appropriate bonding methods of stone (ibid). Densely carved stone ornamentation, which is a characteristics element of Hindu architecture, may be an imitation of carvings otherwise easily performed in wood. Thus, the structure system and ornamentation in early wooden structures was imitated in rock-cut structures and transferred into cut stone structures later. This explains the densely ornamented, trabeated style of Hindu architecture.

In Islamic architecture, ornamentation was intimately tied to the structure. Brick and stucco were the prime building materials in eastern Islamic regions, including Iran. The expansion into the west demanded a translation of brick and stucco forms into stone (Tabbaa, 2001, p. 137). In reference to the popular use of geometric ornamentation in Islamic architecture, Abdullahin and Embi (2013) write that ornamentation on brick surface was easy if geometric. However, with shift from brick to stone, the use of motif ornamentation was found more

Figure 4.3 Arch-shaped entrance door at Lomas Rishi Cave

Source: British Library Board, 15/09/2017, 'Photograph of the entrance to the Lomas Rishi Cave taken by Alexander E. Caddy in 1895', Date: 1895, Shelfmark: Photo 1003/(44b)

Figure 4.4 Barrel-shaped roof of *chaitya* hall at Ajanta (Cave XXVI)

Source: British Library Board, 15/09/2017, 'Photograph of the interior of the Buddhist chaitya hall, Cave XXVI at Ajanta, taken by Robert Gil', Date: c. 1868–70, Shelfmark: Photo 1000/5(516)

appropriate in the Seljuq period (ibid). The importance of geometry in Islamic architecture is explained in detail in the next section. The use of arches, vaults and domes was common, both in brick and stone, and there was a continuous evolution in structural and ornamental styles.

Even though Islamic architecture is identified with arcuate system, Tabbaa (2001) observes that neither trabeated nor arcuated system could gain popularity in Islamic architecture, which consistently strived to create a distinct system that could include destabilising effects. The shift from brick to stone gave rise to many new structural and aesthetical elements which could achieve the heights of destabilisation and became peculiar characteristics of the Islamic architecture in stone – foliate arches, pendent vaults, *muqarna* vaults, polychrome interlaced spandrels (ibid). The use of stone allowed the effect of suspension in voussoir blocks, which was otherwise not possible in brick (Tabbaa, 2001). However, the popularity of these attractive elements was not wide spread, both in geography and in time, due to the difficulty of their design and construction (ibid). Construction of such elements was an exacting process requiring great deal of collaboration in a highly skilled team of masons and artisans (ibid). It was also an expensive process because a lot of stone was wasted, and it also required the finest limestone (ibid). In India, these elements were used in Mughal monuments, thus showcasing their architectural excellence and opulence, as will be discussed later.

While Hindu constructions adopted trabeated structural system and used beams and columns, the Islamic monuments were more advanced in the use of arcuated structural system and the use of arches, domes and vaults were common (Grover, 1996; Luniya, 1978; Sharma, 2001). The difference in the structural system brought clear differences in the ornamentation as well as the overall physical outlook of buildings, and the Hindu and Muslim monuments are identified with visibly distinct roofs and wall openings – Hindu structures with stepped *shikhara* roofs and linear lintels and beams supporting wall openings and gateways and Muslim buildings with curvilinear dome as roofs and semi-circular arches for openings.

4.5 Islamic architecture in Islamic lands

When Islamic constructions began in India around the twelfth century AD, the architecture in the Islamic world had developed its unique elements which were undergoing continuous evolution. The architectural movements in Western Asia, or, more specifically, Seljukian architecture, had a significant influence on the Indian architecture at the time of introduction of Islamic architecture. Tabbaa (2001) focuses broadly on Iraq, Iran and Syria in the eleventh and twelfth centuries and discusses the changes in the Islamic architectural styles from early to medieval period, by linking it with the revival of Sunnis in the early eleventh century. As per Tabbaa's (2001) findings, these changes were driven by the political and religious conditions of that time. Another major driver was the application of knowledge of geometry into art and architecture (Tabbaa, 2001).

The political motive of Seljuqs was also to build a dynastically distant style from the Fatimids, who were more inclined towards naturalism, and this catalysed the shift from organic to geometric art forms, as observed in calligraphy and vegetal ornamentation (Tabbaa, 2001).

The revival of Sunni sect of Islam in the tenth and eleventh centuries is considered the prime reasons for the emergence of many characteristic elements of the Islamic architecture (Tabbaa, 2001). However, it is important to mention that the generalisation of these elements, as Islamic, is not absolute and differences existed in the architecture practised in non-Sunni regions which were eventually reduced to small geographies (ibid). Tabbaa (2011) supports the possibility that most of these elements, which developed at the time of Sunni revival, had a symbolic meaning in addition to being decorative. The meaningfulness is supported by the fact that a sense of decorum was exercised in the use of these elements and an appropriate form of ornamentation was applied to various objects and architectural elements, in consideration to its place, context and function (Tabbaa, 2001, p. 101). For example, the '*girih* mode'[1] was originally applied on objects which had cultic or symbolic value, for example, Quran frontispieces, portals, *mihrabs*, *minbars* and cenotaphs, and later its use spread to other objects and monuments (ibid). Dense vegetal arabesque was more popularly used for cenotaphs and *mihrabs*, where it might refer to "the garden of the Paradise awaiting the deceased or the observant worshipper" (Tabbaa, 2001, p. 101). The geometric order (with or without vegetal arabesque fillets), which also included star patterns as celestial allusions, referred to the ordered universe, whose atomistic and 'occasionalistic' structure needed sustained intervention of the creator (ibid). Thus, the strength and vigour of geometric patterns are associated with the power and authority, and therefore these decorations were more often used for minarets, *minbars* and even for gateways, doors and door frames, which call for attention to their founders (Tabbaa, 2001, p. 101). Referring to the *girih* mode, Tabbaa (2001) argues that "the simultaneity of (symbolic) association stands at the heart of an ornamental system that entirely consumes the object it covers" (p. 102). The *muqarna* is most commonly applied to portal vaults and domes (ibid). The symbolic meaning of *muqarna* and similar other elements which are identified with gravity-defying (or falling down) features, for example, foliate arches, pendent vaults, joggled voussoirs and polychrome spandrels, is associated with the theory of occasionalism (ibid).

4.5.1 Occasionalism

The occasionalistic theory of 'Ashari' theologians argued that

> the world was composed of atoms and accidents. Accidents could not endure within matter (jawhar) for longer than an instant, but were continuously being changed by God. It follows, then, that the attributes of matter (colour, luminosity, shape, etc.) are transitory accidents that change according to the

> will of God, and that even the preservation of matter – the collocation of its atoms – requires the continuous intervention of God . . . who was continually involved in maintaining its order, balance, and coherence. . . . The muqarna dome was intended as an architectural manifestation of this thoroughly orthodox Islamic concept. In order to represent an occasionalistic view of the world, a fragmented and ephemeral-looking dome was created by applying muqarnas to its entire surface, from transition zone to apex. This procedure creates a comprehensive effect intended to reflect the fragmented, perishable, and transient nature of the universe while alluding to the omnipotence and eternity of God, who can keep this dome from collapsing, just as he can keep the universe from destruction.
>
> (Tabbaa, 2001, p. 133)

Other similar destabilising devices used in Islamic architecture are foliate arches, pendant vaults, joggled voussoirs and polychrome spandrels, which will be explained in the following sections. The primary purpose of these architectural expressions seems twofold: first, to symbolise occasionalistic theory by creating architectural illusion of instability among static design elements; second, to emphasise the technical and geometric expertise of Islamic architecture (Tabbaa, 2001, p. 156).

4.5.2 *Vegetal and geometric ornamentation ('girih' mode)*

The use of interlaced vegetal forms and interlocked geometric shapes and patterns is defined as the *girih* mode, which developed significantly during the eleventh and twelfth centuries (Tabbaa, 2001, p. 77). The term *girih* is applied to both vegetal and geometric patterns. There will be an attempt to discuss the two distinctly and understand the causes that led to the inclusion and development of these elements as characteristics of Islamic architecture.

Vegetal ornamentation

While early Islamic ornamental style is closely linked with vegetal ornamentation that existed in other Asiatic regions like China, it became more characteristically Islamic or arabesque with the rise of Sunni sect of Islam in the tenth and eleventh centuries (Tabbaa, 2001, p. 75). Having said that, it is difficult to trace back the development of vegetal arabesque and the exact point at which it changed looks from classical to more properly Islamic, or abstract (p. 79). Some of the earliest examples of true vegetal arabesque, characterised by an advanced degree of abstraction and interconnection, come from eastern Iran, from late eleventh-century palaces and marble cenotaphs at Ghazna (Tabbaa, 2001). The Ghaznavid architectural ornamentations in vegetal arabesque probably inspired Seljuq artisans working in stucco and brick (Tabbaa, 2001, p. 80).

Geometric ornamentation

It is generally believed that the geometric ornament originally developed in Central Asia under Samanids (Sunni Iranian Empire, 819 to 999 AD) in the second half of the tenth century and was transferred in the second half of the eleventh century to the Ghaznavid architecture in Afghanistan and Seljuq architecture in central Iran (Tabbaa, 2001, p. 78). This style used the building material of brick to create the ornament and is therefore characterised by highly textured brick patterns, commonly known as *hazar baf* (ibid). This gave way to the development of complex geometric ornamentations and *girih* mode and the use of overlaid strapworks in the late eleventh century (ibid). Also, the outline in high relief ribs became popular in the late twelfth century in northwestern Iran and Baghdad (ibid).

After Seljuq, the second artistic movement was initiated by Mamluks (1250–1517 AD) in Cairo. The stable government and growing economy during the Mamluk period encouraged competition among architects to use more complex geometrical patterns, particularly in North Africa and Islamic Spain (Abdullahin & Embi, 2013; Crane, 1993). However, limited follow-up of this trend is observed on the Mughal architecture in India (1526–1737 AD). Even though some of the rarest and most complex geometric patterns were used by Mughals, for example, 14-point geometrical pattern on the piers of main dome of Humayun's Tomb in Delhi, the extent was limited and relatively simpler patterns dominated the design in Indian subcontinent (ibid). Abdullahin and Embi (2013) attribute this to "the passion of Indian artisans for symmetrical designs and their insistence on covering all exterior surfaces with ornaments", which was difficult to achieve with the use of complex patterns (p. 250).

Advancement of geometrical science in Iran, Middle East and Central Asia in the eighth and ninth centuries was adopted in practical form in geometrical decorations of buildings in Islamic architecture (Abdullahin & Embi, 2013). Vegetal and floral patterns, derived from Sassanid and Byzantine architecture, were common in Islamic buildings in the seventh and eighth centuries (ibid). The Great Mosque of Kairouan (Tunisia), rebuilt in 836 AD, shows the earliest attempt of applying geometrical ornamentation to Islamic buildings, although vegetal and floral motifs were still the primary decorative elements in the mosque (ibid). The Mosque of Ibn Tulun, built between 876–879 AD, sets the milestone for the use of woven geometrical patterns, and by the end of the ninth century, geometrical patterns became popular among Muslim architects and artisans (ibid). At the time of entering India, Islamic buildings were characterised by Seljuq architecture (1038–1194 AD), which placed a huge emphasis on transforming floral and figural decorations into geometrical patterns. These patterns were often integrated with structural elements, as seen in the dome of Great Mosque of Isfahan in Iran (ibid). The decorations were achieved by the use of carved stucco, wood, coloured glasses, polychromatic tiles, lattice, coloured stone and also tiles and brickwork of geometrical motifs (Abdullahin and Embi, 2013). Alongside geometric patterns, calligraphic inscriptions and vegetal decorations continued to be in use as decorative elements (ibid), although in geometrically reorganised fashion (Kleiss, 2011).

4.5.3 *Calligraphy and monumental inscriptions*

While calligraphic writing has always invited attention in Islam, the development of calligraphy in the eleventh century was characterised by its stricter adoption to geometric orders. The shift from the organic spiral writing style (of Fatimids) to the standardised linear format of writing of Sunnis is associated with their differences in theological tenets and political ideologies. For example, the adoption of strict geometric forms, as opposed to organic spiral writing, is linked with the strict relationship of the state and the masses, or strict political regulations and controls (Tabbaa, 2001, p. 42). It was considered important to use a clear and legible style of calligraphy so as to explain Sunni ideas about the nature of Quran without causing any confusion or ambiguity (Tabbaa, 2001, p. 53). This was also considered politically important to make a distinct calligraphic style identifiable with Sunnis (ibid).

Standardisation of monumental inscription is associated with the intention of "making the word of God or the statement of a dynasty unambiguous and intelligible to all literate people" (Tabbaa, 2001, p. 53). Interestingly, the change in calligraphy style was in parallel to the development of other architectural elements which were tuning into geometric orders, including interlaced vegetal forms, geometric strapworks, and *muqarnas* (Tabbaa, 2001, p. 47). With some reservation Tabbaa (2001) expresses the possibility that the *girih* mode was first used in manuscript (or rather Quranic) illuminations and was later transmitted to architectural ornamentation (p. 48). Having said that, the monumental inscriptions were short on Quranic messages and mostly included political and historical proclamations (ibid). Without getting specific about 'decorative' or 'symbolic' uses of monumental inscriptions, Tabbaa (2001) explains that these were public and official statements that served the following purposes (i) to declare the contemporary concerns of the dynasty that commissioned it; (ii) in an aniconic artistic culture the use of monumental inscriptions was an influential visual means to express religious beliefs; and (iii) while construction of gates, minarets and domes were among other ways of political expressions, the use of monumental inscriptions was the chief means of portraying the political and religious messages in dynastically distinctive manner (Tabbaa, 2001, p. 54). Tabbaa (2001) also presents Necipoglu's and Grabar's reasoning as per which monumental inscription were used to explain the architectural and decorative forms. However, Tabbaa (2001) does not support this explanation fully because to some extent this diminishes the value of un-epigraphic monuments like the Humayun's Tomb in Delhi.

4.5.4 Muqarna

Muqarnas are oversailing courses of small niche sections, formed like honeycombs, often used for interior decoration of half-domes, domes and squinches and at times used in the façade (Kleiss, 2011). It is one of the most original and ubiquitous element of Islamic architecture which originally developed under Seljuq architecture, and its use spanned from central to western Islamic regions between the eleventh and fifteenth centuries. As mentioned earlier, decorative elements were often integrated with structural elements, and *muqarna* is another such element of structural and aesthetical importance. Tabbaa (2001) suggests that the development of *muqarna* in an atmosphere of high religious dogmatism and political and theological shifts from Fatimid to Sunni hints that *muqarna* is also a "symbolic manifestation of an occasionalistic universe and a distinctive emblem of the resurgent Abbasid state, the safeguard of Muslim community" (p. 134). Similar to the purpose of proportioned scripts, the *muqarna* was intended to pay homage to the Abbasid caliphate and encompass the cosmological beliefs of the Sunnis (Tabbaa, 2001, p. 136).

The development of *muqarna* and other elements such as foliate arches, pendant voussoirs and pendant vaults, which practice gravity-defying aesthetics, is associated with occasionalistic theory and Gods-sustained intervention, as explained earlier (Tabbaa, 2001).

4.5.5 *Polychrome spandrels*

Polychrome spandrels or stone interlaces are another "destabilising" element first created by Aleppoian artisans (Tabbaa, 2001). Used mostly over *mihrabs* and portals, the polychrome spandrels might have developed upon joggled voussoirs which are used for lintels and at times for arches in Syrian architecture (ibid). Joggled voussoirs and polychrome spandrels use two or more colour stones, cut in angular or curvilinear shapes which interlock into each other to create a plane surface. Although in later examples, this effect was achieved through stone inlays. The use of these elements was not very prevalent, neither in the Islamic world nor in India. In the Indian context, weak resemblance to joggled voussoirs may be seen in the arch of the Alai Darwaza. Polychrome stone inlays are relatively more commonly seen in Indo-Islamic architecture; however, it was applied on other elements in addition to spandrels, for example, in Humayun's Tomb in Delhi, and was also not necessarily geometric, for example, in the Taj Mahal. Thus, the structural purpose, in the Indian context, is fully defeated, and these were merely decorative elements.

Discussions above help us in understanding the theories guiding Hindu (although briefly) and Islamic architecture in Delhi. It can be concluded that at the time of the first interaction of Hindu and Muslim society, their structural and ornamentation styles were a direct antithesis to each other. Negotiation and innovation followed next, and the Islamic architecture in India reached its climax in the sixteenth century. Stage-wise progression of Islamic architecture in India is discussed in the next two sections.

4.6 First phase of Islamic architecture – the Delhi Sultanate

By the twelfth century, the architectural vocabulary of Islamic buildings was well defined, and most of the elements were borrowed from the Roman and Byzantine systems (Sharma, 2001).

Figure 4.5 Corbelled arches of the screen at Quwwat-ul Islam Mosque

Source: Photograph by Harsh Tripathi, 2017

Figure 4.6 Corbelled dome at Quwwat-ul Islam Mosque

Source: Photograph by Harsh Tripathi, 2017

Figure 4.7 Corbelled arches at Sultan Ghari Tomb

Source: Photograph by Harsh Tripathi, 2017

The change of structure system from trabeated to arcuated was gradual, and most of the initial Islamic buildings are adorned with arches and domes, which are built using the trabeate system. The tendency of the artisan to mimic the structural system from the past into new materials without realising the appropriateness was clearly visible in the built form. As mentioned earlier, in the first stage of mimicry, during the Hindu period, wooden prototypes were directly copied in stone. In this second stage of imitation, the momentum of first stage was continuing and beams and columns of stone were used to create circular shapes, which are otherwise built using stone wedges or voussoirs, basic module of arcuated structural system. This technique gave birth to corbelled arches and domes, as seen in the screen of arches (or *riwaq*) (Figure 4.8) of the Quwwat-ul Islam Mosque (Figures 4.5 and 4.6), the arches and vaults of the Sultan Garhi Tomb (Figure 4.7) and other contemporary buildings of the Slave dynasty. The Tomb of Balban (last ruler in the league of Slave dynasty) was among later constructions of Slave dynasty, and this for the first time used true arches, built from radiating voussoirs (Sharma, 2001).

4.6.1 *Quwwat-ul Islam Mosque*, c. 1193[2]

Quwwat-ul Islam Mosque is the first work of Muslim builders in India, although the building cannot be taken as an ideal example of true Muslim style of construction (Beglar, Carlleyle, & Cunningham, 1874). It was a period of constant warfare for the Muslim rulers, who were not very well received in India. However, the faithful needed the mosque as a place of worship and also as a symbol of expression of the launch of Islamic rule (Grover, 1996). There was not enough time to prepare fresh building materials, and therefore, readymade materials were removed from existing temples in Quila Rai Pithora, probably 27 Hindu temples, including a few Jain temples (ibid). Thus, the mosque was constructed in a rush by Hindu masons under Muslim supervision. Making best use of their knowledge of trabeated (or post and lintel) structural system, Hindu masons constructed Islamic building elements – arches and domes. Thus emerged "trabeated" (or corbelled of rather untrue) arches and domes, which are good examples of fusion arising from the interaction of foreign supervisors with native artists.

Beglar, Carlleyle and Cunningham (1874) claim that the mosque stands at the site of the main Hindu temple of the old Hindu city of Delhi. This is demonstrated by the fact that tall pillars of the Hindu temple remain undisturbed (have single Hindu shafts) and unaltered from their original location (ibid) (Figure 4.9). On the contrary, the colonnade around the mosque (Figure 4.10) was built by Muslim builders by reassembling building materials raided from different Hindu temples or probably from multiple chapels of the main temple (Beglar, Carlleyle & Cunningham, 1874). The colonnade is a "faulty" reassembly of mismatching Hindu elements put together in haste (ibid). The incongruity goes unnoticed due to eye-catching ornamentation on the pillars, but when observed closely the ornamentation itself is deformed (Figure 4.11). This is so because the Sunni sect of Islam strongly disapproved of figurative representation, and therefore, human forms on Hindu pillars were disfigured by cutting face or facial parts (Grover, 1996). From their archaeological observations, Beglar, Carlleyle, and Cunningham (1874) conclude that

> this great beautiful structure is essentially Hindu in design, altered to a greater or lesser extent by the Muslim conquerors . . . who . . . deliberately did their best to hide the signs of the Hindu origin of the structure by building in, covering up, whitewashing and plastering, destroying parts and building them up according to their own crude and barbarous notions, and crowded the whole (structure) by inserting in the true style of oriental exaggeration in their inscription, that they built the structure.
>
> (p. 45)

Figure 4.8 Quwwat-ul Islam Mosque

Source: Recreated by authors with base image from Murray (1911)

Figure 4.9 *Qibla* of Quwwat-ul Islam Mosque at the site of a former Hindu temple, as seen from behind the arched screen

Source: Photograph by Harsh Tripathi, 2017

Figure 4.10 Pillars from different Hindu temples used to create the cloister around Quwwat-ul Islam Mosque

Source: Author, 2016

Figure 4.11 Deformed human figures on the columns from a Hindu temple, used inside Quwwat-ul Islam Mosque

Source: Photograph by Harsh Tripathi, 2017

Very often, spoliation by early Muslim rulers is interpreted as a politico-military approach for imposition of Islam on native Hindus (Patel, 2004). Patel (2004) challenges this belief and argues that spoliation was practised in India even before, whereas it was not the case in the Islamic world. Instead, "Ghurid governors of northern India extensively patronised local building practices, which had originated in the architectural traditions of the recently annexed region" (Patel, 2004, p. 37).

As mentioned in earlier chapters, there were stages of cultural, political and religious mergers and the shifts were evident in the architectural style as well. The buildings in the first phase are often seen as demonstrations of initial tensions, both political and religious, between the native Hindus and the Muslims (Welch, 1993). The usual argument is that in addition to conquering land and acquiring wealth, the Muslim rulers were also under religious obligation to convert maximum population to Islam (Welch, 1993). In the process of Islamisation, the practical and influential architectural language was chosen as the medium of communication, and the physical surroundings were appropriately modified to confirm with Islamic style to the extent possible (ibid). Thus, many Hindu temples were modified and building elements were rearranged and deformed to suit the functional and visual requirements of an Islamic place of worship, to the extent possible (Grover, 1996). Mosques, so constructed, were serving the dual purpose of a place for worship for the faithful as well as the symbol of change of the political control (Patel, 2004). Often, this spoliation is interpreted as tantamount to the destruction of Hindu religion and Indic worship (Patel, 2004), attributing to the popular concept of transformation of India from Dar al-Harb (not a territory of Islam) to Dar ul-Islam (territory of Islam) (Welch, 1993). However, Patel (2004) challenges this notion and claims that spoliation was already in practice in India since the first century AD, as observed in the case of Buddhist stupa being reused as Jain stupa. Also, the use of recycled building materials was not uncommon in Hindu temple architecture although this practice is less documented (Patel, 2004). Patel (2004) concludes from historic records, that "the measures taken by the Ghurids in northern India were negotiations with, rather than eradications of, the local traditions and infrastructure" (p. 37).

In the architectural world, the change of style, from Hindu to Islamic, was achieved gradually and witnessed various stages of transition during which many new architectural elements were born out of Indo-Islamic fusion. This initial phase of Islamic building construction showcases the reluctantly adapted Hindu style, more as a compromise rather than a thoughtful merger. However, this marked the beginning of the Indo-Islamic architectural style that achieved excellence during the Mughal rule, which is observed as one of the richest periods in the architectural history of India.

Most of the buildings of this phase, characterised by rearrangement, reshaping and reuse, are located along Delhi and Agra, and the only two examples outside this zone are Adhai Din Ka Jhonpra and the Jama Masjid at Bari Khatu (Nagpur) (Sharma, 2001).

4.6.2 *Qutub Minar*

With some surety, the historians claim that Qutub Minar is truly Islamic in its purpose and it follows the then popular practice of attaching a minaret (or *Mazinah* or *Muazzin*'s tower) to the mosque, as also seen in other contemporary mosques like Ibu Tulun (876 AD) and Sultan Barkut (1149 AD) in Cairo and *minar* at Koel (1252 AD) (Beglar, Carlleyle & Cunningham, 1874). Another proposition is that it was originally planned as a tomb tower by Aibak, similar to Kishmar in Khurasan, but later Iltutmish added three more storeys and converted it into a victory tower (Merklinger, 2005). Merklinger (2005) concludes that there is uncertainty regarding the function of Qutub minar to be funerary or symbolic, "but there is no doubt that the immense structure proclaimed the might of Islam to the whole world" (p. 20).

The rich style of ornamentation of lower three stories has at times confused archaeologists, some of whom doubt that these lower stories were constructed by Hindu ruler Prithviraj Chauhan. However, Alexander Cunningham clarifies that the rich work of the balconies in the lower towers is much in accordance with the rich traceries of the early mosques in Delhi and Ajmer, which were built by Hindu masons under Muslim builders' supervision and therefore the continuum of Hindu-style ornamentation is observed (Beglar, Carlleyle &

Cunningham, 1874). The usual belief is that the momentum of rich ornamentation of the Hindu style was difficult to avoid because early Muslim conquerors were depending upon native Hindu masons in carrying out their designs. It is important to raise another contrasting probability of these stalactite-like pendentives being truly Islamic *muqarnas* (Peck, 2005), which were already a very popular ornamental and iconographic element of Islamic architecture outside India, as explained in an earlier section (Figures 4.12 and 4.13). This is further supported by the downward-looking concave units filled with geometric and inscriptional ornamentation. Further research on the matter is necessary before any conclusion can be drawn.

The last two stories were later constructions by Firuz Shah Tughlaq, who restored and rebuilt the fourth floor by replacing it with two floors (Merklinger,

Figure 4.12 Rich ornamentation on the arched screen of Quwwat-ul Islam Mosque

Source: Photograph by Harsh Tripathi, 2017

Figure 4.13 Rich ornamentation on balconies of Qutub Minar

Source: Photograph by Harsh Tripathi, 2017

2005). The difference in style of ornamentation, construction and material of the lower three towers and the upper two towers is attributed to the time difference and change of architectural taste which had taken place over a century time lapse between Quṭb al-Dīn Aibak and Firuz Shah Tughlaq (Beglar, Carlleyle, & Cunningham, 1874). Often the combined use of two materials for last two stories is associated with Firuz Shah's liking for these materials (ibid). However, this may be contended, for the economic condition of that time that did not allow the use of rich materials and also restricted the ornamentation as demonstrated by other monuments constructed by Firuz Shah, discussed later. It could well be the case that Firuz Shah tried to reassemble what was originally created by Iltutmish, and the use of fresh material would have been minimalistic.

Figure 4.14 Rich ornamentation inside the Tomb of Iltutmish

Source: Photograph by Harsh Tripathi, 2017

Despite the fact that Indian artisans lacked skill in Islamic designs, the buildings inside Qutub complex are adorned in *girih* mode and calligraphic inscriptions, as seen on the screens of Quwwat-ul Islam Mosque, Qutub Minar and inside the Tomb of Iltutmish (1211 – 1236 AD) (Figure 4.14). There is a strange mix of imperfectness and richness in the ornamentation of these buildings, thus reflecting the seriousness in effort in imposing the excellence of Islamic world architecture on this newly acquired territory, which lacked training in these designs.

This was an era of continuous development in architecture, and the first true arch was constructed at the Tomb of Balban, which is otherwise an unattractive building belonging to the Balban dynasty (1266 and 1287 AD) (Brown, 1956). Brown (1956) interprets this advancement in building construction as a symbol of socio-political progression of Islamic rule, under which foreign artisans were introducing the produce and knowledge of their land. "Delhi was by this time becoming a city of repute, of wealth and influence, a centre of attraction to men of distinction, culture, and learning, possessed of wide scholarship, practical knowledge, and technical skills" (Brown, 1956, p. 15). Except for the 'House of Balban', there was minimal building activity for the next three-quarters of a century after the decline of the Slave dynasty in 1234 AD (ibid).

4.7 Khilji dynasty (1290–1320 AD)

With the rise of Afghan Turks, Alauddin Khilji ascended the throne of Delhi in 1296 AD. This was the time when the Seljukian Empire in West Asia was dissolved by Mongol invasions, thus causing the out-migration of Seljukian artisans and craftsmen, many of whom found refuge in India under the Khiljis (Brown, 1956). The association of Khilji buildings with Western Asia is therefore noticeable, and the contribution of expert Seljukian masters from the west is evident in the sudden rise in the quality and maturity of building art, as seen in the Alai Darwaza (Figures 4.15 and 4.16).

The Alai Darwaza built in 1305 by Alauddin Khilji, in Qutub complex is one of the four planned entrances to the Quwwat-ul Islam Mosque. Although the structural and ornamental maturity of this gate might not be a perfect match for the Seljukian monuments in Western Asia, it still was a huge step up from earlier Islamic constructions in Delhi. One of the most attractive features of this structure is the use of red sandstone and marble together in the façade, decorated in arabesque. While the pointed horseshoe-shaped arch of Alai Darwaza is not only uncommon in the Islamic land where it originated, it became a unique element of the Khilji architecture in India. The underside of the arches is ornamented with lotus buds, which are typical of Seljuk-style floral ornamentation, the details of which can be found in Abdullahin and Embi (2015) (Figure 4.17). Grover (1996) suggests the weak possibility of this decoration being inspired from Jain *torana*. As discussed earlier, the use of joggled voussoirs, symbolic of 'occasionalistic' theory, was not uncommon in Islamic architecture, and the lotus buds carving on the underside of the voussoirs could probably be a new version of joggled voussoirs. Occasionalism is also demonstrated by downward-looking spearheads circumscribing the arch and seems to be a weak copy of stone pendant voussoirs. Another typical element of Islamic architecture on Alai Darwaza is the carved spandrel carrying abstract vegetal ornamentation of Seljukian style. Brown (1956) emphasises upon the architectural importance of Alai Darwaza for introducing the architectural elements that were reproduced, with appropriate modifications, in the styles that followed. In general, the key characteristics of Khilji architecture, most of which are found in Alai Darwaza, include the use of alternate courses of headers and stretchers; pointed horseshoe-shaped arches; broad domes; squinches with recessed arches (or *muqarna*); geometric perforations on windows; decorative moldings; arabesque low reliefs; and the use of sandstone relieved by marble (Sharma, 2001).

Along similar lines of construction other building activities were undertaken at the time of Alauddin Khilji, including Alai Minar, which remains incomplete till date (Brown, 1956) (Figure 4.18). Another important mention is the establishment of the city of 'Siri' in 1303 (Blake, 1991). This was the second capital city in Delhi, after Qila Rai Pithora, but the first to be founded by a Muslim ruler. However, not much of it is remaining (ibid). To the west of 'Siri' was an interesting water body called Hauz-e-Alai (modern-day Hauz Khas), which was probably a place of

Figure 4.15 Alai Darwaza, with Qutub Minar in the background

Source: Photograph by Harsh Tripathi, 2017

Figure 4.16 Alai Darwaza, from the side
Source: Photograph by Harsh Tripathi, 2017

Figure 4.17 Lotus bud carving on the arch of Alai Darwaza
Source: Photograph by Harsh Tripathi, 2017

Figure 4.18 Incomplete construction of Alai Minar
Source: Author, 2016

recreation for the Khiljis in the hot weather of Delhi (Grover, 1996) and might have also contained small villages of the modest people who were not residing within the walls of Siri. Later construction by Khiljis, in and around Delhi, for example, Ukha Masjid in Bharatpur, Rajputana, show gradual distancing from the high standards set by Alai Darwaza (ibid). This is attributed to the political and economic weakening of the Khilji dynasty, which met its end in 1320 AD. (Brown, 1956).

4.8 Tughlaq dynasty (1320–1413 AD)

At the time when the Tughlaqs took over the Delhi Sultanate in 1320 AD, Central and Western Asia were already a part of Mongol Empire, and the chances of their invasion into India were very high. Perhaps the preparedness for invasion was reflected in the design of buildings which were constructed by the Tughlaqs (Brown, 1956). For example, the stone masonry was laid in buttressed fashion such that the thickness of the wall resided on top and the outer façade was inclined, usually at 75-degree angle (ibid). This was the technique used for constructing stronger walls in brick masonry in Multan (modern-day Pakistan) and Punjab, and the same was imported to Delhi as a technique to build strong buildings (ibid). Many construction projects in India and Multan were simultaneously undertaken by the Tughlaqs, and therefore the imitation in style was obvious

(Brown, 1956). Circular bastions were provided at almost all intersections of the walls of forts, palaces as well as in mosques and tombs (ibid) (Figure 4.20).

Even though this style of wall construction directly symbolises strength and preparedness for war, its use for tombs and mosques was conflicting with their purpose and the associated calmness and peacefulness (Brown, 1956). In the context of the tomb of Ghiyath al-Din Tughlaq, Brown (1956) suggests possible explanations for the fort-like structure of the tomb: (1) the tomb was connected with the citadel and was designed in a fashion that it can take the form of a self-contained fortress, like a miniature barbican or the outpost to the city; or (2) it could have been a *donjon* or place of last resort.

Ghiyath al-Din Tughlaq (1320–25), the first of the Tughlaq dynasty, founded the fortress city of Tughlaqabad, the third capital city in Delhi. It was strategically located at the highest point of the rock, thus hinting at the effort towards safety from the constant threat of attack from foreign and native adversaries (Brown, 1956). The design and construction techniques were guided by safety and defence mechanism. The peripheral walls are constructed using massive rubble stone masonry cladded with dressed quartzite stones available on the site (ibid). Double-storey bastions were provided at regular intervals. On the inner side of the wall run vaulted corridors (Figure 4.21) with arrow slits for short and long range. The fort (east), city (north) and palace (west) are put next to each other, thus fulfilling the residential and military requirements together. Brown (1956) suggests that these complexes were designed after the Roman fashion, where the citadel and the city were laid adjoining to each other. This was also a common format of development of the tenth-century European and Western Asian cities and pre-Islamic Indian cities, including Qila Rai Pithora. The stepwell or *baoli* was another Hindu city component as observed in Qila Rai Pithora that continued to be in use in Tughlaqabad, although its shape changed from circle to square. This could have been the influence of linear designs of Islamic planners. This is also evident in the grid iron pattern of streets of the city (Figure 4.19). Despite nearness of Delhi to the River Yamuna, which could have easily served water requirements for any of these new cities if only these were positioned along the bank, these cities were instead located at the highest points of the ridge. Perhaps safety was a much bigger concern, as political tussles were frequent.

Along similar lines of design and construction, two other cities, namely Jahanpanah[3] (meaning world's refuge) and Firozabad,[4] were founded, respectively, by Muhammad bin Tughlaq (1325–51) and Firuz Shah Tughlaq (1351–88).

The tomb of Ghiyath al-Din Tughlaq continued to use only a few elements from the Khilji dynasty, including the 'spearheads' in the arches (Brown, 1956). It also introduced new features, like the Tudor arch, and also its joint use with lintel beam (ibid). The Tudor arch was imported from Central Asia, and probably Indian masons lacked confidence in the new shape and used an additional lintel beam along with the arch, to doubly ensure the structural stability (ibid). This is an uncommon example of fusion of trabeate and arcuate systems. The structural value of lintel beam was slowly lost to its ornamental appeal, and the arch and lintel combination continued to be the architectural style of future buildings

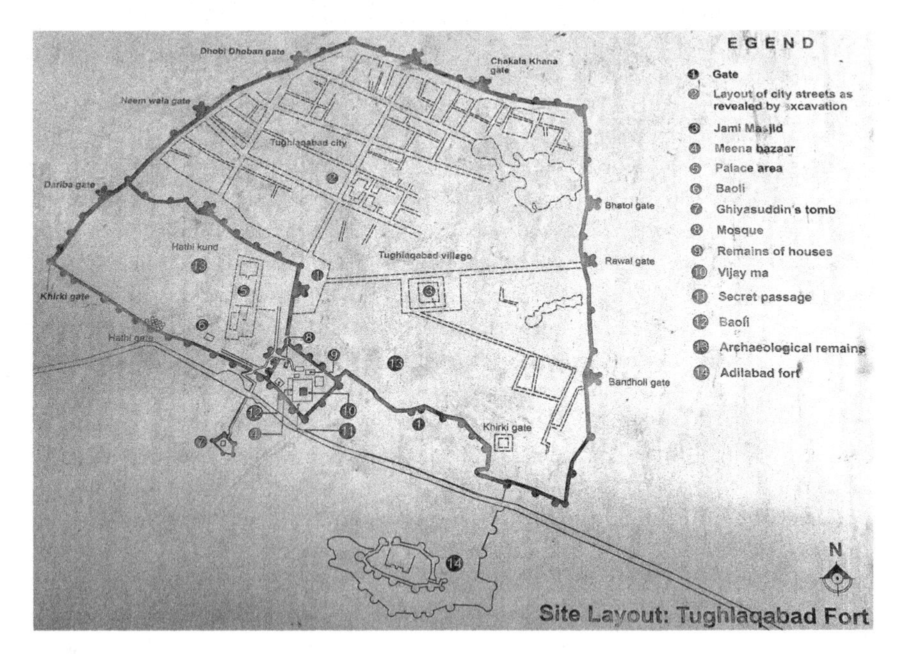

Figure 4.19 Layout of Tughlaqabad Fort City

Source: Author, 2016

Figure 4.20 Bastions on the walls of Tughlaqabad Fort

Source: Photograph by Harsh Tripathi, 2017

Figure 4.21 Vaulted corridors running along the peripheral walls of Tughlaqabad

Source: Photograph by Harsh Tripathi, 2017

Figure 4.22 Rampart (this page) connecting Tughlaqabad Fort with the tomb of Ghiyath al-Din Tughluq (next page)

Source: Photograph by Harsh Tripathi, 2017

Figure 4.22 (Continued)

Figure 4.23 Stepwell in Tughlaqabad

Source: Author, 2016

(ibid). The tomb is mostly built in red sandstone, with certain portions, including the dome, in white marble. In the dearth of Islamic invention of dome finial, the tomb of Ghiyath al-Din Tughlaq is crowned by a finial resembling the *kalasa* or *amla* of a Hindu Temple (Grover, 1996). In summary, the architectural style of Ghiyath al-Din Tughlaq puts special emphasis on building strength, as compared to ornamentation. His buildings are characterised by the use of inclined walls with bastions, the Tudor arch, the combined use of the arch and lintel, spear-heads on fringes and the dome finial of Hindu *kalasa* style (Brown, 1956).

Later, around 1326–27, his son Muhammad bin Tughlaq created the city of Jahanpanah by enclosing the area between earlier cities of Qila Rai Pithora and Siri. This became the fourth capital city in Delhi. The writings of Ibn Battuta, who lived at Delhi between 1333 and 1340, suggest that Muhammad intended to include Tughlaqabad as well, thus forming one big city by combining all the existing cities of his time, but perhaps the expense was too huge (Peck, 2005). With some reservations, Peck (2005) proposes that the old city of Lal Kot was probably the urban area and Siri was the military zone, while the remaining area in-between was reserved for the royal palace. There are very limited physical remains of this city, and therefore the architectural style is unknown. A part of the original wall still exists as *Satpula* (meaning seven bridges) (Peck, 2005) (Figure 4.24). The structure is a dam with sluice gates and reservoir and served the dual purpose of storing water for irrigation and ensuring safety for the city. It has seven arcuated openings, which give the structure its name.

Muhammad is known for his extraordinary and unpopular schemes that were executed with intentions to expand the empire. One such step was to shift the capital and its entire population from Delhi to Daulatabad, located 775 miles south of Delhi, in Maharashtra. The purpose was twofold, first, to establish a strong Muslim base in Deccan (or South India) and, second, to allow the buildup of a large military force in Delhi for a campaign in Afghanistan (Peck, 2005). Due to various reasons, his plans failed and brought catastrophic political and economic consequences. In the process, the royal treasury was exhausted, and skilled artisans and masons were dispersed from Delhi.

Figure 4.24 Satpula dam

Source: Author, 2016

It was under these conditions when Firuz Shah ascended to the throne after Muhammad's death in 1351 (Peck, 2005). Despite Firuz Shah Tughlaq's passion for building art, the buildings of his time are characterised by "the use of inexpensive material, put together in readiest manner, in a plain but serviceable style" (Brown, 1956, p. 22). Mostly quartzite stone from local quarries was the primary building material, and the use of imported stones, like red sandstone and marble, was discontinued, even for cladding and ornamentation. Ornamentation, though rare, was in the form of moulding in plaster rather than the carving on stone (Brown, 1956). In addition to that, a shift was made from finished stone ashlar to unfinished rubble stone masonry with plaster (ibid). Roughly dressed monoliths were used for lintels, door-posts and pillars alike (ibid). Aforesaid depreciations in the quality of construction are attributed to the poor economic condition and dearth of skilled artisans in Delhi caused by temporary relocation of capital to Daulatabad (ibid).

At the time when Firuz Shah ascended to the throne, in 1351, he chose to operate from Jahanpanah until 1354 (Peck, 2005). Later he founded a new, fifth capital city of Firozabad, closer to the banks of the River Yamuna, as against the trend of locating on high ridges, thus indicating the start of a peaceful time and architectural developments. During these years, the population inside the three contiguous walled cities of Qila Rai Pithora (first city), Jahanpanah (fourth city) and Siri (second city) was also growing, as implied by the large size and growing number of mosques constructed in Jahanpanah by Firuz Shah, like Begumpuri Masjid (c. 1370) and Khirki Mosque (1375) (Peck, 2005). The construction of a large *madrasa* at Hauz Khas also supports this argument. In addition to that, other settlements were developing around religious centres like the 'sufi' shrines in Mehrauli and Nizamuddin. However, the distant city of Tughlaqabad (third city) was to some extent abandoned, at least by the court (Peck, 2005).

The fortress palace at Firozabad, called Firoz Shah Kotla, was the trendsetter for the palace forts in the Mughal time. Even though the design was possibly a reproduction of the castled palaces in Rome and Byzantine, it was the first to lay down the principles for the palace forts in India and to be developed later by the Mughals, as seen in the Red Fort of Shahjahanabad. Within the palace complex is located a pyramidal structure with three receding stories that host a famous Ashokan pillar (Figure 4.26). Emulating the act of Quṭb al-Dīn Aibak when he installed the Iron Pillar at the courtyard of Quwwat-ul Islam Mosque, Firuz Shah relocated two famous Ashokan pillars from Ambala, one of which was installed into his palace.

Noticeable contributions were made in the design of mosque and tomb erected during Firuz Shah's time. In the absence of surface ornamentation, the grandeur of the mosque was kept up by raising the whole structure on a substructure of arcuated cells (or *tahkhana*) and adding a long flight of steps around it, of which Jami Masjid, built around 1354 at Firozabad, inside Firoz Shah Kotla is the first example (Merklinger, 2005). The cells at the ground floor were presumably used as *waqf* property for the purpose of *serai* or rental shops to help the upkeep of the mosque (Merklinger, 2005). Creating ground-floor properties could also be seen as a financial strategy to support religious activities at the time when the economy

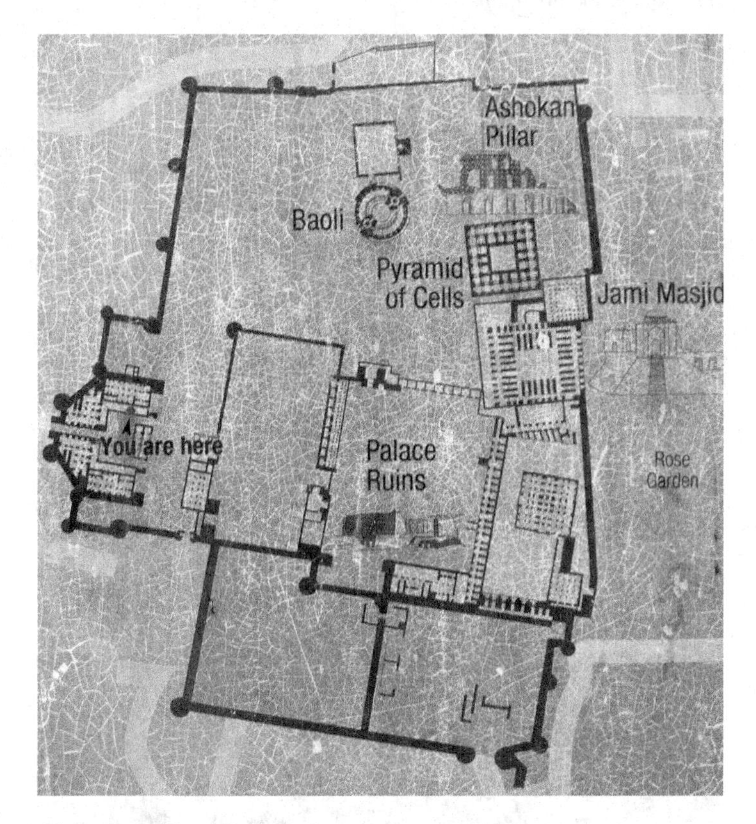

Figure 4.25 Layout of Firoz Shah Kotla

Source: Photograph by Harsh Tripathi, 2017

Figure 4.26 Pyramidal structure with Ashokan pillar at Firoz Shah Kotla

Source: Photograph by Harsh Tripathi, 2017

Figure 4.27 Jami Masjid at Firoz Shah Kotla (with Firoz Shah Kotla Cricket Stadium in the background)

Source: Photograph by Harsh Tripathi, 2017

Figure 4.28 Circular stepwell at Firoz Shah Kotla

Source: Photograph by Harsh Tripathi, 2017

was sluggishly recovering from the financial crisis of capital's shift to Daulatabad. As an architectural style, the concept of raising the mosque on *tehkhana* gained an immense popularity and was followed in the design of many important mosques, including the Jama Masjid at Shahjahanabad.

Another noticeable variation is observed in the plan of Khirki Mosque (at Jahanpanah) and Kali Mosque (at Nizamuddin). It is believed that Khirki mosque was built before Firuz Shah founded Firozabad and was in fact his contribution to Jahanpanah city, where he lived between 1351 and 1354. Kali Mosque is a

Figure 4.29 Khirki Mosque at Jahanpanah; from outside (top) and inside (bottom)

Source: Photograph by Harsh Tripathi, 2017

small-scale version of the same. In both the mosques, the open-air courtyard is sub-divided into four smaller courtyards by run of cross-axial colonnaded bays with multiple small domes on the roof (Merklinger, 2005). Entrances on all three sides are flanked by tapering turrets built in rubble masonry and plaster, typical of Tughlaq style (ibid).

In reference to tomb architecture, it is usually believed that the octagonal tomb of Firuz Shah's important official Khan-i-Jahan Tilangani, built after his death in

Figure 4.30 Tomb of Tilangani

Source: Photograph by Harsh Tripathi, 2017

1368, was the first octagonal tomb in the Islamic history of India (Brown, 1956; Merklinger, 2005). The tomb has battlemented walls which are surrounded by a varandah pierced by arches. Merklinger (2005) is of the view that the design was copied from the Tomb of Zafar Khan which was built inside the tomb complex of Ghiyath al-Din Tughlaq. Another view suggests that the octagonal tomb takes inspiration from the Tomb of Shah Rukh-e-Alam in Multan (Brown, 1956). Ghiyath al-Din Tughlaq had constructed this around 1320–24, for his dynasty but later gifted to a 'sufi' saint after whom it is named (Department of Archaeology and Museums, 2004). However, the Tomb of Tilangani is believed to be the model for numerous octagonal tombs in the future. Located close to the Nizamuddin Dargah in modern-day Nizamuddin area, the tomb is devoid of its historical importance and is rather engulfed into the organic outgrowth of native residential settlements around it.

Octagonal tomb design was given much more importance by the Lodis who developed it further, as will be discussed later. However, Firuz Shah's own tomb, built in advance by himself, has a square plan and avoids the use of turrets at wall intersections while continuing with the use of battlement walls (Brown, 1956). The tomb is located next to Hauz-e-Alai tank, which was also repaired by Firuz Shah and renamed as Hauz Khas, as it is known today (Peck, 2005). Firuz Shah also added a mosque and a *madrasa* (or college of theology) at Hauz Khas complex (Grover, 1996).

In summary, Firuz Shah Tughlaq made significant contributions to architecture, of which austerity was the style as well as the necessity of the time.

4.9 The Sayyid (1414–51), Lodi (1451–1526) and Suri (1539–45)[5]

The end of the fourteenth century is marked by destructive loots by Taimur in India, and the capital city of Delhi was brought to its lowest point in all respects – social, economic and political (Merklinger, 2005). The century after that was a period of cultural, social and architectural revival and "an interlude, an appendix to the Delhi Sultanate as well as a preface to the Mughal period" (Merklinger, 2005, p. 47).

Due to the dearth of money and artisans, all of which were taken away by Timur, it was challenging to put up any work of architectural significance, and Delhi was devoid of any major construction for at least a decade after Taimur's departure (Merklinger, 2005). Following the footsteps of his predecessors, the first Sayyid ruler Khizr Khan established a new capital city in 1418, called Khizrabad (ibid). Later, in 1433, his successor Mubarak Shah Sayyid started but never completed the city of Mubarakabad (ibid). The sequence of finding new capital cities in Delhi was finally put to a halt, and no new cities were built by any other ruler of this dynasty (ibid). The second Lodi ruler, Sikandar Lodi moved the capital to Agra and founded Sikandarabad (ibid). As the fifteenth century moved on, a new architectural style, which combined many Hindu and Muslim elements, started emerging on royal tombs, which were the predominant building type during this time (ibid).

Unlike earlier dynasties, which built tombs only for royalties and saints, the Lodis are known for creating tombs for other important courtiers and noblemen (Brown, 1956; Merklinger, 2005). In Merklinger's (2005) view, this practice was due to the changing attitude towards kinship. Lodis, who were a settled Afghan tribe, believed into their tribal tradition as per which the ruler was merely the first among the equal others (Merklinger, 2005). The treatment of equality towards other fellow beings resulted in a significant increase in the number of tombs in Delhi during this period. However, the distinction between the ruler and others was maintained in the layout of tomb, and octagonal plan was reserved for the ruler, while other tombs were square in plan (Brown, 1956; Merklinger, 2005), although the square-shaped tomb of Bahlul Lodi, the founder of the Lodi dynasty, is an exception.

Within Delhi, the three octagonal tombs of this time period are identified with three rulers[6] – Mubarak Shah Sayyid (1421–34), Muhammad Shah (1434–45) and Sikandar Lodi (1489–1517). Starting with Tilangani, the trend of octagonal tomb culminated in the splendid tomb of Sher Shah Suri built in Sassaram, Bihar (Merklinger, 2005). Like in the earlier period, the octagonal tomb was surrounded by a veranda with arched openings (Merklinger, 2005). The veranda and projections above its arches were probably an outcome of the protection required to the plaster covering the rubble masonry (ibid). Even though the use of tapering turrets was discontinued, its effect was not fully removed and the vertex of octagonal veranda has an additional attachment, much like the Gothic buttress, inclined at an angle (Brown, 1956). This is even more noticeable among straight vertical lines of the walls and columns. The octagonal tombs were single-story structures with one central dome and a pillared domed kiosk on each side. Not repeating the mistake from Tilangani's tomb (Figure 4.30), visual corrections were made to improve the vision of the domes when seen from the ground level. The central dome was raised on an octagonal drum, and other secondary domes were raised on pillars to form the kiosks (or *chatri*, meaning umbrella). To add emphasis to the central dome, ornamental pinnacles (or *guldasta*) were added at vertices of the octagon. These corrections were first performed at the Tomb of Mubarak Shah Sayyid (1432) (Figure 4.31) and were repeated, with further refinements, at the Tomb of Muhammad Shah Sayyid (1444) (Merklinger, 2005) (Figure 4.32).

Despite attempts to improve the sight of central dome, its vision was masked by the *chatris* in front. The third octagonal tomb of Sikandar Lodi (1517) (Figure 4.33) has avoided the use of *chatri* altogether and is more neatly finished than the earlier two. While carrying forward the turrets at its corners, the tomb of Sikandar Lodi has many improvements in design, most important ones being the ornamental gateway on the south, the mosque in the west and the walled, elevated garden surrounding all these structures (Brown, 1956; Merklinger, 2005). The tomb marks the transition between the fortified walled compounds of the Tughlaqs and the extensive garden tombs of the Mughals that followed (Brown, 1956).

As mentioned earlier, many square tombs were constructed in the honour of important courtiers, some of which are even larger and more imposing than the royal octagonal tombs (Brown, 1956). A broader conclusion is drawn by Brown (1956) who finds that octagonal tombs are one-third larger in plan than square tombs, while in elevation, the latter is one-third taller than the former (ibid). The tall height of walls of square tombs was hidden by the use of blind arcades, as also seen at Alai Darwaza. In Delhi, there are approximately seven square tombs of this order, most of which remain unidentified, as they do not bear the name of those they commemorate (ibid). These are isolated structures without walled enclosures. Made in rubble stone masonry and plaster, the walls were built straight, unlike before. The façade was designed to have a central rectangular portion projecting out of the wall and holding a large, recessed arch reaching up to the parapet. The arch contains a small trabeated doorway of Hindu style, with ornamented brackets on either side. A single pointed dome on squinches roofs the entire structure.

During the time of the Sayyids and Lodis, no major public mosque was constructed, and instead small, private mosques were built as an attachment to the tomb, of which the earliest example is the mosque attached to the Bara Gumbad (Brown, 1956) (Figure 4.35). Later examples of private mosques were built during the time when Delhi was officially under Humayun's rule (1530–40-interregnum-1555–56) (ibid).

After the grand victory over the Lodis in 1526, Babur (1526–30) became the first Mughal emperor of India. While Babur mostly operated from Agra, when his son Humayun took charge, he decided to move his residence from Agra to Delhi. In 1533, Humayun chose the site of an old earthen citadel and refurbished it into a fortified city of 'Dinpanah' (meaning asylum of faith), today known as the

Figure 4.31 Tomb of Mubarak Shah Sayyid

Source: Photograph by Harsh Tripathi, 2017

Figure 4.32 Tomb of Muhammad Shah Sayyid

Source: Photograph by Harsh Tripathi, 2017

Figure 4.33 Tomb of Sikandar Lodi

Source: Photograph by Harsh Tripathi, 2017

'Purana Qila' or Old Fort. At this starting phase of the Mughal rule in India, there was serious struggle to maintain political power which constrained architectural projects of any significance to happen. During the lean time of the Mughal rule, between 1540 and 1555, Sur dynasty took control of Delhi and installed themselves at Dinpanah, which was probably extended and renamed as 'Shergarh' by

Figure 4.34 Bara Gumbad (top) and Shish Gumbad (bottom)

Source: Photograph by Harsh Tripathi, 2017

Figure 4.35 Mosque attached to the Bara Gumbad (top); interior of the mosque (bottom)

Source: Photograph by Harsh Tripathi, 2017

Figure 4.36 Qila-i-Kuhna Mosque (front façade)

Source: Photograph by Harsh Tripathi, 2017

Sher Shah Sur in 1540. The two cities together are called the sixth city of Delhi. The foothold of Mughals was weakly established in the start, and despite the political change of power in 1526, the legacy of Lodi architecture continued for nearly half a century longer (Brown, 1956).

The highest architectural achievement of Sher Shah's time was made in the Qila-i-Kuhna Mosque at Purana Qila, built in 1541 (Figures 4.36, 4.37, and 4.38). The continuity of turrets was maintained although in a subdued fashion, as seen at the corners of the central arched opening. Turrets lost the functional role and

Figure 4.37 Qila-i-Kuhna Mosque from inside

Source: Photograph by Harsh Tripathi, 2017

Figure 4.37 (Continued)

were reduced to ornamental elements. At the rear façade, which is much less ornamented, the size and ornamentation of the circular bastions is much more extensive. Small cantilevered openings or *jharokhas* of Hindu style, supported on cornices ornamented with geometrically organised floral carvings of Islamic order, are provided over arched openings as well as on the side and rear façade. There was a significant improvement in the quality of ornamentation, and the use of *muqarna* is noticeable in the vaults of the mosque. Combined use of white marble and red and yellow sandstone is seen the first time under Lodi-style buildings. With high-quality material and workmanship, the monument is considered a

Figure 4.38 Qila-i-Kuhna Mosque at 'Purana Qila' – side and rear views

Source: Author, 2016; Harsh Tripathi, 2017

"connecting link between the old style of Imperial Delhi and the oncoming style of the Mughals" (Brown, 1956, p. 29), which will be discussed in the next section.

4.10 Mughal architecture (1526–1858)

With the return of Humayun in 1555, the Mughal rule was re-established in India, this time much more firmly footed to last for next three centuries. After the accidental death of Humayun, Akbar (1556–1605) ascended to the throne, and this marked the real start of a new empire as well as the beginning of golden era in Mughal architecture in India, which was taken to an unmatched level of excellence by his descendant Shah Jahan (1628–58). Discussions on Mughal architecture are often focused on richness of material and visual attractiveness of ornamentation consequential to the progression made in the political and economic domains. In addition to that, it is necessary to recapture the discussions from Chapter 1 and bring into the discussion the religious and political doctrines adopted by Mughals, which became an important guide to the development of Mughal architecture.

Alam (1997) writes that between the twelfth and fifteenth centuries, during pre-Mughal Islamic rule in India, the religious laws also guided worldly affairs, and there were limitations in addressing political complexities concerning multi-religious society of the Indian subcontinent. The integration of non-Islamic systems, like Sassanid state system, into the Muslim world was discouraged to the extent of considering it a religious sin (Alam, 1997). This significantly limited the scope for political adjustments under the Islamic rule, and the situation was such that the primacy of religious laws was acknowledged but political matters were decided on the basis of expediency (Alam, 1997). Gradually, Islamic political theorists allowed, to varying degrees, the integration of non-Muslim institutions into the political body of Islam (ibid). For example, Sikandar Lodi encouraged non-Muslims to learn Persian and serve in the state management (ibid). At the time of non-Muslim Mongol invasion of Islamic regions in the thirteenth century, the edifice of Islamic culture was shaken (ibid). Under these situations, the political theory of Persian scholar Khwaja Nasir al-Din al-Tusi, written in his book *Akhlaq-e Naseri* published in 1235, gained a lot of importance as the new Islamic political ideology (ibid). Per this theory, it was possible for an ideal city to be constituted by people from multiple religions, and the role of an ideal ruler was to ensure the well-being of his subjects from all sects (ibid). This, in a way, confirmed religious tolerance and political integration of non-Muslim institutions, and therefore the approach of Mughals, since inception, was much more accommodative of non-Muslims native formats prevailing in India at that time. This philosophy of harmonious integration with natives helped the Mughals achieve a successful state system under which economic and political achievements were plenty. The political, social and religious integration of the Mughals was an influential strategy that helped them in establishing symbiotic relationships with the existing kingdoms, most of which eventually became a part of the Mughal Empire. Thus the empire quickly expanded its boundaries and stretched into the east to include Bengal and Gujarat in the west. Alongside opening opportunities for sea trade, the provinces of Bengal and Gujarat also gave exposure to the rich architectural styles that developed over time into these regions. Excessive use of marble in Mughal monuments may well be attributed to the inclusion of Rajput state of Jodhpur with its rich marble quarries of Markrana, which were earlier the exclusive asset of Rajput rulers. Subsequent evolution of political and religious ideologies of Mughals probably gave way to admissibility of many non-Islamic architectural elements into the Islamic architecture. Mughal architecture innovations have major borrowings from Timurid style from homeland of these rulers which was appropriately amalgamated with Indian as well as European styles (Asher, 1992).

Timurid architecture excelled in using complex geometrical formulae-based floor plans and a highly sophisticated arcuated system of walls that allowed covering large, open spaces by a narrower superstructure (Asher, 1992). The sophisticated technique of replacing solid walls with interconnected and stacked transverse arches (Figure 4.44) resulted in the creation of structures with a large central room surrounded by small chambers, accessed through these interconnected arches (ibid). In mature stages of Timurid architecture, the eight smaller chambers around the central tomb chamber of a mausoleum became symbolic of eight levels of paradise (ibid) (Figure 4.45). Such layouts and techniques, popularly used in Samarkand mausoleum (Gur-e-Amir), were inherited by Mughals and first used in India for Humayun's Tomb (ibid). The Timurid layout continued to inspire later tombs built under the Mughals (ibid). An element of distinction in Mughal architecture was the use of geometrical patterns in floor finishing (Abdullahin & Embi, 2013), which might have developed in response to geometrical floor plans (Figure 4.42).

Even though Timurid architects excelled in garden development, the combination of tomb and garden, as used in Humayun's Tomb (Figure 4.41), was

Figure 4.39 Humayun's Tomb (front elevation)

Source: Photograph by Harsh Tripathi, 2017

Figure 4.40 Half-dome at Humayun's Tomb

Source: Photograph by Harsh Tripathi, 2017

Figure 4.41 Charbagh garden around Humayun's Tomb

Source: Photograph by Harsh Tripathi, 2017

Figure 4.42 Geometric pattern on floor inside Humayun's Tomb

Source: Photograph by Harsh Tripathi, 2017

unknown in Central Asia and Iran and seems to have been inspired from Lodi tombs in India. Other commonalities with Sultanate architecture are the use of a sub-structure or *tehkhana* for raising the monument, polychrome façade of red sandstone, white and black marble, *chatris* decorated in polychromatic tiles and *guldastas* (Michell & Pasricha, 2011). A noteworthy element of Timurid architecture was the use of a double dome in which the outer dome extends beyond the drum and takes a bulbous form, which terminates into a pointed tip.

Figure 4.43 Eight-point geometric pattern on stone lattice *mihrab* inside Humayun's Tomb

Source: Photograph by Harsh Tripathi, 2017

Large gardens with water channels were symbolic of paradise and formed an essential part of important Timurid cities (Michell & Pasricha, 2011). Even though this concept relied on stepped terraces built on hillsides in Samarkand, it was replicated with a lot of effort on the plains of Agra and Delhi (ibid). Mughals were keen to imitate the garden settings of their homeland, and this eventually

Figure 4.44 Transverse arches inside Humayun's Tomb, connecting the central room with the adjoining cells

Source: Photograph by Harsh Tripathi, 2017

resulted in a highly developed garden layout around tombs and palaces, for example, in the Taj Mahal at Agra and at Red Fort in Delhi. This also gave way to an advanced water channel system inside palatial buildings, for both aesthetic and cooling purposes.

In Akbar's time, the use of marble was prevalent as complimentary decorative material to red sandstone, as seen at Humayun's Tomb (Figure 4.39). In 1562, Akbar's nobleman, Farid al-Khan, rebuilt the shrine (dargah) of Nizamuddin Auliya (Asher, 1992). The four walls of the tomb were built as marble lattice (or 'jail') supported on marble pillars decorated in intricately carved geometric patterns (ibid) (Figure 4.46).

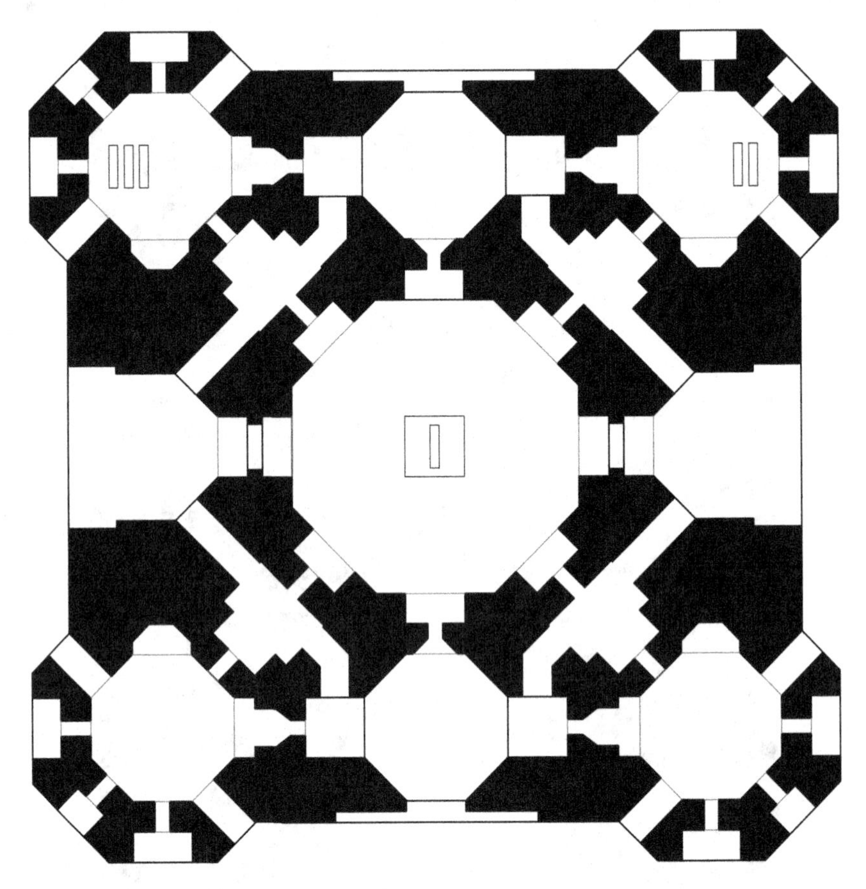

Figure 4.45 Layout of Humayun's Tomb

Source: Recreated by authors with base image from Murray (1911)

Figure 4.46 Marble lattice (or *jali*) walls and marble columns at Nizamuddin Auliya Dargah

Source: Photograph by Harsh Tripathi, 2017

Asher (1992) writes that due to the use of marble in this tomb, the material became an "emblem of sanctity" in Mughal architecture (p. 42). This can explain the marble construction of later tombs and mosques, like the Taj Mahal at Agra and Moti Masjid at Delhi. In addition to sacred monuments, the use of marble was extended to royal buildings during Shah Jahan's time.

The Mughal architecture reached its highest point of excellence and opulence under Shah Jahan regime. His architectural contributions to tombs, palaces, hunting pavilions, gardens and planned cities are considered extraordinary even by modern standards (Asher, 1992). The wealth and power of the emperor facilitated rich architectural expression in the form of increased scale and elaboration of construction project and extensive use of marble cladding (Michell & Pasricha, 2011). The highly precise execution of intricate architectural elements of Persian origin, like the *muqarna* domes and vaults, foliate arches and bulbous domes, is demonstrated in Shah Jahan's constructions. The sandstone columns of Hindu temple style, which were popular at the time of Akbar, were replaced by Italian-style columns with tapering, fluted shafts, bulbous, leafy bases and brackets as bunches of buds (Figure 4.49). Another popular intake from Italy was the technique of *pietra dura* (Figures 4.50 and 4.51), which is discussed later. This was the first time when European architecture was showing a strong presence in the Indian subcontinent, thus signifying growing international linkages. Most popular examples of Shah Jahan's rich architectural style are the Taj Mahal in Agra and the Red Fort at Delhi.

Agra was the epicentre of the Mughal Empire until 1648, when Shah Jahan decided to shift the capital back to Delhi (Michell & Pasricha, 2011) and built the seventh and the last Islamic capital city of Delhi, which was named after him as 'Shahjahanabad'. The grandeur of this new capital city surpassed Agra and Lahore which were earlier the most important Mughal cities (Asher, 1992). In 1637, Shah Jahan felt that at Agra and Lahore, the space was inadequate for royal processions and court ceremonies, and, therefore, he demanded a bigger capital

Figure 4.47 Red Fort, Delhi

Source: Photograph by Harsh Tripathi, 2017

Figure 4.48 Chatta Chowk bazaar (top) and Naqqar Khana (bottom) inside Red Fort

Source: Photograph by Harsh Tripathi, 2017

in Delhi, the former nucleus of Islamic dynasties in India (Asher, 1992). The layouts of the city and palace were planned by Ustad Hamid and Ustad Ahmed, and foundations were laid in 1639 (ibid). With additional assistance from other architects Aqil Khan and Aqa Yusuf, the construction of the Red Fort was completed in April 1648 (ibid).

The fort is laid out in an irregular pentagon plan, essentially rectangular in shape, covering an area of 125 acres circumscribed by walls of red sandstone, 3 km in perimeter. Among many royal monuments of architectural importance is the marble throne of Shah Jahan (Figure 4.49) installed in the Hall of Public Audience (Diwan-i-Am). The Lahore Gate of the fort connects the Hall of Public Audience with the city, through a 70-m-long stretch of bazaar with two-story arcades which serve as shops even today (Asher, 1992). The concept of a covered bazaar travelled from Iran to India and is very well executed in the city of Shahjahanabad. The Chatta Chowk (or four-square) breaks the monotonicity of the long array of shops and forms a small gathering square, typical of Iranian bazaars (Figure 4.48).

The bazaar ends into the 'Naqqar Khana' or Drum Room, through which was announced the arrival of the emperor (ibid). In the hall, the throne of the emperor is positioned at the end of the central bay. It is built in white marble, thus signifying the semi-divine role of the emperor. Four balusters support the deeply sloping *bangala* roof, adopted from provincial architecture of the Bengal region. The marble throne, as well as the marble wall behind it, is decorated with inlays of precious and rare stones, a technique known as *pietra dura*, which developed in Florence in the sixteenth and seventeenth centuries. Trade relations with Europe were maturing at this time, and thus the import of architectural styles and materials, like the black marble inlaid tiles used on the walls behind thrones, was also starting to happen. This hints at growing international linkages of the Mughals and also their increased admissibility of non-Islamic architectural elements.

The shift towards naturalistic ornamentation including birds, animals and floral motifs under the Mughal architecture was also suggesting that religious and political ideologies of Islam, which guided building ornamentation, had evolved over time. As discussed earlier in this chapter, the early forms of Islamic ornamentation were semi-naturalistic until Sunni revival in the medieval period (tenth-twelfth century) when geometric order was applied to all art forms, including vegetal ornamentation, calligraphy and monumental inscriptions (Tabbaa, 2001). Systematic ornamentation is also associated with the strict control and order of the government. The evolution of religious and political ideologies of Islam was visible in return to naturalistic vegetal ornamentation under Mughal architecture. It is believed that natural vegetal ornamentation is devoid of symbolic meaning, except that the gardens are referred to as paradise (Department of Islamic Art, 2001). Thus, its use on the throne emphasises the semi-divine persona of the emperor.

While the Hall of Public Audience is located in the centre of the Red Fort, buildings reserved for private use of the emperor, like the white marble riverfront pavilions, are aligned along the river along the eastern wall. A water channel, called Nahr-i Behisht, meaning Canal of Paradise, runs across these private buildings and shows up at ornamented marble pools, specially created inside private chambers (Figure 4.52).

The luxury and comfort of Mughal emperors is demonstrated in the use of *hammam* (or bath), which was a special addition to Indian building types. The bath is located near the Hall of Private Audience and complemented its function of holding private meetings on the most important state matters with the team of a

Figure 4.49 Diwan-i-Am (Hall of Public Audience, top); marble throne of Shah Jahan (bottom), inside Diwan-i-Am, Red Fort

Source: Photograph by Harsh Tripathi, 2017

Figure 4.50 Diwan-i-Khas, Red Fort

Source: Photograph by Harsh Tripathi, 2017

Figure 4.51 *Pietra dura* at Diwan-i-Khas, Red Fort

Source: Photograph by Harsh Tripathi, 2017

Figure 4.52 Nahr-i Behisht and marble pool inside Rang Mahal, Red Fort

Source: Photograph by Harsh Tripathi, 2017

selected few (Asher, 1992). The environment of the bath was conditioned to suit the summer and winter temperature at Delhi (ibid).

The wealth and pomp of Shahajahanabad extended beyond the palace walls into the city where members of the royal family and high-ranking ladies constructed markets, mosques and gardens, while other noblemen built mansions mirroring the form and functions of the royal palace (Asher, 1992). In addition to that, imperial women like Akbarabadi Begum (first wife of Shah Jahan) and Roshan Ara (youngest daughter of Shah Jahan) also developed suburbs around Shahjahanabad (Asher, 1992). These include the Shalimar Bagh and Roshanara Bagh and tomb, which still form a part of the green network and serve as district-level parks in Delhi (ibid).

Figure 4.53 Entrance to Jama Masjid (top); prayer hall (bottom), Shahjahanabad

Source: Photograph by Harsh Tripathi, 2017; author, 2013

The city of Shahjahanabad was dotted by many small and big mosques commissioned by imperial women. It is interesting to note the gradual increase in the size and number of mosques that were built in the city, thus indicative of the growing size of population. In 1653, the city expanded over an area of 6,400 acres, holding a population of 400,000 people (Asher, 1992, p. 200) (Figure 4.53). The last in the series was the Jama Masjid, built by Shah Jahan in 1656, which is one of the largest congregational mosques in Indian subcontinent to date (Asher, 1992).

Among the bazaars, the most famous was Chandni Chowk (or Moonlit Street) laid by the eldest daughter of Shah Jahan, Begum Sahiba Jahanara (Figure 4.54).

Figure 4.54 Chandni Chowk bazaar in 1815 (top) and 2017 (bottom), viewed from the Red Fort

Sources: British Library Board, 15/09/2017, Watercolour of Chandni Chowk in Delhi from 'Views by Seeta Ram from Delhi to Tughlikabad Vol. VII', Date: 1815; Shelfmark: Add.Or.4827; Harsh Tripathi, 2017

The name of the market is derived from the moon light that reflected into the water of the canal and lit the market at night. This was the main commercial thoroughfare connecting the Lahore Gate of the Red Fort in the east to the Fatehpuri Masjid in the west. This layout was in line with the Iran cities, where the bazaar connected the palace to the main mosque, thus symbolising the arterial connection of the head (palace) to the heart (mosque), as mentioned earlier in Chapter 3. The written corpus of travellers Manucci and Bernier describe Chandni Chowk

Figure 4.55 Begum Samru's Palace in 1858 (top) and in 2017 (bottom), Shahjahanabad

Source: British Library Board, 15/09/2017, 'Part of a portfolio of photographs taken in 1858 by Major Robert Christopher Tytler and his wife', Date: 1858; Shelfmark: Photo 193/(12); Harsh Tripathi, 2017

as a vibrant bazaar laid out along a water canal lined with trees (Asher, 1992). Harshness of weather was combated by pillared galleries in front of shops, similar to that found in Chatta Chowk inside the Red Fort. To the north of Chandni Chowk was the Sahibabad garden with a large *seria* to host rich merchants who will come here for trade from all parts of the world.

Four major urban elements of this city were the mosque, citadel, bazaar and residential quarters of rich merchants and courtiers. While these were exclusive in function, their interaction and interdependencies were consequential of maturing society, polity and economy. A wealthy group of specialised service providers, like Begum Samru and Chunamal, was developing outside the walls of the royal fort, thus giving a noteworthy architectural expression to the city. Samru's Palace is now used as a branch of the Central Bank, although the building is poorly maintained,

Figure 4.56 House of Chunamal, Shahjahanabad

Source: Photograph by Harsh Tripathi, 2017

like many other surrounding structures, together adding to the dense, mixed used character of Chandni Chowk (Figure 4.55). However, the house of Chunamal is now occupied by his successors and is relatively well maintained (Figure 4.56).

Mirza Ghalib ki Haveli is a good example of contemporary residential structures of the Mughal time period. The house is located in dense, narrow by-lanes of Chandni Chowk and is restored as a museum, dedicated to Ghalib (Figure 4.57).

After achieving the unmatched heights of excellence during Shah Jahan's regime, Mughal architectural style of that period continued to guide all later constructions of the Mughals. In the rich corpus of Mughal monuments in Delhi, the Tomb of

Figure 4.57 Ghalib ki Haveli, Shahjahanabad

Source: Photograph by Harsh Tripathi, 2017

Figure 4.58 Tomb of Safdarjung

Source: Photograph by Harsh Tripathi, 2017

Figure 4.59 Interiors of the Tomb of Safdarjung

Source: Photograph by Harsh Tripathi, 2017

Safdarjung (1753–54) is probably a modest construction by the Nawab of Awadh Province, Shuja-ud-Daulah in 1753–54 (Figure 4.58). Despite the use of rich elements of design, like the *muqarna* half dome at the entrance gateway, the material used in the tomb is contrastingly modest, signifying the state of destitution and downfall of the Mughal dynasty (Figures 4.59 and 4.60). An important structure in the complex is the mosque, which is topped with three bulbous domes with their ribs

Figure 4.60 Front façade of the Tomb of Safdarjung

Source: Photograph by Harsh Tripathi, 2017

Figure 4.61 Mosque attached to the Tomb of Safdarjung

Source: Photograph by Harsh Tripathi, 2017

highlighted in red (Figure 4.61). Although the tomb draws attention, it is probably not the most exquisite of the Mughal constructions.

By the time the Mughal Empire reached its end in the mid-nineteenth century, Mughal architecture was established as the prototypical Indian building mode, adopted by Hindu and Muslim patrons as well as by the British for their civic projects (Michell & Pasricha, 2011). Limited adoption of Mughal architecture is noted outside India, for example, in the Royal Pavilion at Brighton built in 1787 (ibid). After a long phase of political struggle, the Mughal rule officially ended in 1858, and the British Raj was founded (Asher, 1992). Consequential to the change of empire was the slow death of this royal city, which was transformed from a seat of government into a commercial and industrial hub, as it exists today. The overgrowth of many informal industries and mixed-use developments in Shahjahanabad have gradually overshadowed the grandeur it once boasted of.

4.11 The British Raj (1858–1945) and European architecture in Delhi

Architecture is a non-verbal expression of the cultural, religious and political ideologies of the society in correspondence with its economic ability. The era of Islamic architecture in India is an excellent example of such expressions. The shift of political power from Mughals to the British was definite to bring changes into the built environment, in a way that buildings display the political aspirations of this new protagonist. Architecture is also the vocabulary that communicates the relationship of the state and the people, or the attitude of the government. The objective behind British colonisation of India was to export raw material from India and import finished goods (Lang, Desai & Desai, 1997). The profit-maximising objectives of British government kept them at a distance from harnessing the local art, craftsmanship and architecture (ibid). The British rule was characterised by intellectual and political isolationism, as per which they insisted upon cultural, architectural and political superiority over Indians (ibid). As a response to this, the spirit of nationalism was flared among Indians, who started looking upon their pre-colonial past as a source of pride and inspiration. Nationalism, as a response to isolationism, was executed in the architectural world as a movement against 'modernism', which was a synonym for 'Westernisation' in India. Lang et al. (1997) identify three major forces, post-British rule, which impacted the urban design and architecture in India: first, the Industrial Revolution; second, the British colonisation and country's reaction to it; and, third, the internal tussle, post-independence, for political and intellectual hegemony. The following section discusses the influence of such socio-economic and political movements, alongside architectural movements in Europe and America, on the built environment of Delhi.

It is important to mention about the socio-economic situations in Britain and its influence on the architectural world. This was the time when Britain was industrialising and the process of mechanisation resulted in human degradation (Lang, Desai, & Desai, 1997). The response of architectural world was the

Arts and Crafts movement, which grew out of Gothic Revival and advocated the continuation of the role of craftsmanship by maintaining the tradition of decorative arts (ibid). The works of John Ruskin (1819–1900) influenced many others, like William Morris (1834–96), Edward Burne Jones (1837–98) and Philip Webb (1831–1915), who in turn influenced the Indian architecture (ibid). The Arts and Crafts movement was a broader social movement to mitigate the negative influence of the Industrial Revolution and safeguard the existence of local craftsmen (ibid). However, the philosophy to harness Indian craftsmanship was in a way conflicting with the political ideology of British architects in India who ardently supported political control of the British on India. The Arts and Crafts movement was differently approached in India, and European designs, mainly of Gothic order, were executed with the help of local craftsmen. To some extent, it supported local craftsmen but not local craft. Also, the division of labour was explicitly defined, and local craftsmen were merely the executors of European designs created by British architects.

Political ideology of British architects also influenced the nature of architectural education in India (Lang, Desai, & Desai, 1997). For example, the Sir JJ School of Arts, founded in 1857 in Mumbai under the leadership of British teachers, mandated teaching European art to Indian students while appropriately omitting Oriental and Indian art (ibid). This approach was criticised by Lockwood Kipling, who founded the Mayo College of Arts in Lahore with the intention to develop Indian art and craftsmanship to the extent that even the building was erected by a local craftsman, Bhai Ram Singh (Lang, Desai, & Desai, 1997, p. 88). In summary, there were two different schools of thoughts at that time, one favouring the import of European architecture in India, and another believing in the application of traditional Indian architecture to contemporary designs. Very few belonged to the latter school of thought, which was later developed into 'Indo-Saracenic' architecture and ironically became the official architectural style of the British in India, as discussed later (Lang, Desai, & Desai, 1997). Saracenic was a term used by Christians for the Islamic world and its architecture (ibid). Hence, the Islamic architecture in India was referred as Indian Saracenic architecture (ibid). However, 'Indo-Saracenic' architecture may be better understood as a style that emerged from the attempts of British architects to use Indian and Saracenic elements in buildings of essentially European character. The symbolism in Indo-Saracenic architecture is interpreted in many ways, for example, as a symbol of political dominance of British over Mughals, as a demonstration of the British to show their sense of belonging to India and as a healer to exploitive nature of British imperialism (ibid). The Indo-Saracenic movement began in 1870, and, amid other competing styles, it lasted till Independence and even later, although with varying levels of fusion.

4.11.1 Shifting the capital from Kolkata to Delhi

For the majority period of British rule in India, Kolkata was their capital city, and thus most of the developments of British architecture were concentrated here. Kolkata was also the city where the nationalist movement originated and reached at its peak in 1911, when Kolkata became an inhospitable home for the British. The political tensions in the city grew to the extent of relocation of the British capital from Kolkata to Delhi. However, the built environment of Delhi was dominated by Mughal impressions and was not updated to the same extent as in Kolkata, to represent the political aspirations and aesthetic taste of the British. It took another 20 years for the architects to build a new capital city in Delhi for the British, known as New Delhi. This is the eighth and the last capital city in Delhi to date.

The very earliest European settlements of the late eighteenth century were located on the southern side of the fort, in Daryaganj (Peck, 2005) (refer Figure 3.4). From the time of taking control of Delhi, in 1803 till the mutiny of 1857, the British officials were residing in the northeastern section of the walled city, near Kashmiri Gate, while the army was stationed around Daryaganj (ibid). During this time, many European-style buildings were constructed inside Shahjahanabad, including St. James Church (1836) (Figure 4.63) and residential bungalows of British officials, like Ludlow Castle (ibid) (Figure 4.62). After the mutiny, the British were more cautious of maintaining distance, both socially and physically, from the native Indians (ibid). This was strictly manifested in the walled city, and immediate changes were observed inside Shahjahanabad, to the extent that the prayers at Fatehpuri Masjid and Jama Masjid were discontinued, and the former was instead sold off to a Hindu merchant and the latter was used as a ballroom (ibid). The

Figure 4.62 Ludlow Castle in 1878, Civil Lines

Source: British Library Board, 15/09/2017, 'Photograph taken by an unknown photographer in c. 1878, of Ludlow Castle in Delhi', Date: 1878, Shelfmark: Photo 752/16(23)

Figure 4.63 St. James Church in 1858 in Shahjahanabad, an example of early European architecture in Delhi

Source: Photograph by Harsh Tripathi, 2017

European community gradually moved out of the walled city into the Civil Lines area while the military moved inside, occupying the Red Fort and Daryaganj (ibid).

During the time between 1911, when the formal announcement for shifting the capital to Delhi was made, and 1931, when New Delhi was built, the Civil Line served as the temporary capital for the British (Peck, 2005). Later, the British offices and families relocated to New Delhi, and many vacant colonial buildings in Civil Lines were acquired by Delhi University, and are still in use (ibid). European bungalows in the area were occupied by wealthy Indian families, mostly those who embraced a Western lifestyle (ibid). To date, the area is dotted by European-style buildings, of which the Mutiny Memorial (1870), designed as per Victorian Gothic architecture, is worth mentioning (Figure 4.64). A major challenge for the British was to overrule the existing identity of Delhi, as the royal seat of Mughals. As a part of this exercise, and also as a profit-making venture, the role of Shahjahanabad was changed to serve as a commercial hub, as discussed in Chapter 3. Bias against Old Delhi was visible, as the development strategies were transforming its image from the imperial capital city to a commercial hub crowded by commoners, as it exists today. This doubly enhanced the importance of New Delhi as the capital city for the exclusive use of the government and its officials, who were mostly British.

This environment of political hegemony was also cascading into the architectural world. Most architectural firms in the country were run by foreign architects, majority British, many of whom had spent their entire life in India (Lang, 2002). While most of the Indian architects were foreign educated, Sir JJ School (founded in 1857) in Mumbai was the only school of architectural education in

Figure 4.64 Mutiny Memorial (1870), Civil Lines, an example of Gothic architecture (Art and Craft movement)

Source: Photograph by Harsh Tripathi, 2017

India and was primarily led by British-trained architects (ibid). They harnessed the shift towards modern architecture in cognisance to the Indian context (ibid). With the presence of majority British- and foreign-trained architects, the influence of international architectural styles on India was evident. During the first half of the twentieth century, the architecture in India was a mix of multiple streams of thoughts, of which Gothic Revival, Art Deco, Neo-Classicism and Modernist architecture made noteworthy impressions.

4.11.2 Gothic Revival and Art Deco

In the first half of the twentieth century, two distinct international styles were visible in parallel in the Indian architecture – the Gothic Revival and the Art Deco (Shah, 2008). The Gothic Revival movement that emerged in England in mid-eighteenth century spread into Europe and America throughout the nineteenth and twentieth

Figure 4.65 Garrison Church (1930), an example of Gothic Revival

Source: Photograph by Harsh Tripathi, 2017`

centuries (ibid). Delhi, being under the British Empire, was certainly witnessing the style as seen in the Garrison Church (1930) (Figure 4.65) built by Arthur Gordon Shoosmith (1888–1974) and St. Stephen College (1941) (Figure 4.66) built by Walter Sykes George (1881–1962) (ibid). This style was identified with the use of stone and brick masonry although the ornamentation was unfaithful to the original medieval Gothic architecture. On the contrary, Art Deco was anti-historical and highly individualistic in nature and also used ornamentation elements (ibid). It is characterised by undulating asymmetrical forms, which became even more feasible after the invention of reinforced concrete in France in 1898 (Shah, 2008).

The Art Deco or 'modern style' as known in England descended from Art Noveau and influenced Europe between 1890 and 1910 (Lang, 2002). This style was popular in India during 1930s and 1940s and continued till 1960s, by which time it was 'passe' in Europe (Lang, 2002). Architecture in India was growing as a shadow of foreign styles, and the time lag between the two may be attributed to the slow pace and reluctance with which the new styles were accepted in India. Lang (2002) writes that "in a time of political upheaval in India (around 1945) Art Deco was too radical an art form to be acceptable to imperialists and too foreign to be acceptable to Swaraj idealists" (p. 12).

During all these years, Mumbai was the epicentre of architectural thinking in India, and Art Deco expressions are very strongly imprinted on the city, for example, in hotels along Back Bay (modern-day Churchgate) (Lang, 2002). The style also spread to other major cities, including Delhi. Among a few surviving examples of this order are the Imperial Hotel and Golcha Cinema (Figure 4.68). The Imperial Hotel (Figure 4.67) was built in 1931 under the supervision of British architect C. G. Blomfield, who was one of the many architects working alongside Lutyens on the New Delhi project (*The Imperial India*, 2017). The hotel occupies a prime location on Janpath Street, the erstwhile Queensway, which was the second most important social boulevard in the country (the first being the Kingsway, now known as Rajpath) (ibid). The Regal Theatre, built in 1931 and designed by Walter George,

Figure 4.66 St. Stephen College (1941), an example of Gothic Revival

Source: Photograph by Harsh Tripathi, 2017

Figure 4.67 Imperial Hotel (1931), Janpath, an example of Art Deco

Source: Photograph by Harsh Tripathi, 2017

Figure 4.68 Golcha Cinema (1954), Daryaganj, an example of Art Deco

Source: Photograph by Harsh Tripathi, 2017

is another noteworthy example of Art Deco in Connaught Place which was the commercial centre for New Delhi, as it is today. Another, later example of Art Deco in Delhi is by W. M. Namjoshi, who designed the Golcha Cinema in Daryaganj (Golcha Cinema, 2007). It opened in 1954 and is still in use. The theatre has a cement-rendered façade, painted in a pink colour. Similar to Art Deco movie theatres in Mumbai, the façade is imposed with cut-out lettered signage, placed on the top and on the side of the building. With continuing patronage from wealthy clients and the flamboyant use of Art Deco on pompous cinemas and bungalows, Art Deco was interpreted as 'modern' in India.

4.11.3 *Swadeshi architecture as a response to Art Deco (1920s and 1930s)*

There was another nationalist school of thought that questioned the appropriateness of foreign architecture on the grounds of its irresponsiveness towards Indian climate, cultural, history and ethos of nationalism. In addition to that, the foreign-trained Indian architects were well aware of the ferment of European and American architecture during the first three decades of the twentieth century, and this increased their appreciation towards Swadeshi styles, as visible in the nationalist designs of Santiniketan in West Bengal state, east of India by Surendranath Kar (1892–1970) (Lang, 2002). In 1920s and 1930s, while the rest of the world was innovating 'international style', Indian architecture was participating in the Swadeshi movement. It was not until after independence that modern ideas were absorbed into Indian architecture, and for many years, its embracement was limited to a few architects, as will be discussed later (Lang, 2002).

The revivalist movement of Indian architecture was a bold and radical departure from the contemporary styles, both in India and abroad (Lang, 2002). The Swadeshi architecture or 'the modern Indian architecture' movement harnessed the ethos that "the past can give order to the present and be a source of identity and pride for the people who see it as part of their culture" (Lang, 2002, p. 26). Sris Chandra Chatterjee (1873–1966) was the main driver behind this movement.

Figure 4.69 Lakshminarayan temple (1938), an example of Swadeshi architecture

Source: Photograph by Harsh Tripathi, 2017

Figure 4.70 Mahasabha Bhawan (1939), an example of Swadeshi architecture
Source: Photograph by Harsh Tripathi, 2017

This was a reaction to three important architectural developments of that time: first, the use of Art Deco style in India; second, the emergence of 'international modernism' in Europe; and third, the imposition of Western architecture in Japan (ibid). Chatterjee turned to the examples of historic cities and based his designs on Indian canonicals, particularly the *Shilpa Shastras* (Indian science of art and craft). Among a few examples of Chatterjee's works are the Lakshminarayan Temple (Birla Mandir) (1938) (Figure 4.69) and the Mahasabha Bhawan (1939) (Figure 4.70) in Delhi. Despite the resemblance of form with ancient temples, the use of material and functional organisation was kept modern.

4.11.4 *Neo-Classicism in Neo-Delhi – 1911 to 1931*

Given the struggles for political and architectural hegemony between 1911 and 1931, when New Delhi was under construction, the decision on the architectural statement of buildings in this new capital city was very crucial. While buildings were meant to be a statement of European supremacy, they also had to communicate the inclusiveness for native Indians, even if in a limited sense. More so, the rise of nationalism, both in politics and in architecture, was to be addressed strategically without increasing frictions. The answer was found in Neo-Classicism of Sir Edward Lutyens (ibid). Since the style was rooted in traditional European architecture, classicism was considered an appropriate response to nationalism. Typical of Indo-Saracenic technique, elements from the Mughal and Hindu architecture were admitted, to a tolerable extent, in the design of the two most important buildings of New Delhi, the Viceroy's House and the Secretariat buildings. The Viceroy's House, now Rashtrapathi Bhavan, is counted among Sir Edward Lutyens's noteworthy works of Neo-Classical style. Prater (2017) admires the contextualisation in this work, which fuses Lutyens Neo-Classicism with

Mughal architecture, thus blending British and Indian culture. In a balanced way, inspiration was taken from Sanchi Stupa and its railings into the design of the domed roof of Viceroy's House (Peck, 2005). Also he used *chatris*, of Mughal and Sultanate architecture, to ornament the roof. From his works in Delhi, Lutyens borrowed a few ornamentation elements into his later designs outside India; for example, the door cornices for his last residential work, Middleton Park (1934) use bells of Delhi order (Prater, 2017). Along similar lines, Herbert Baker designed the Secretariat buildings, which are essentially of the Neo-Classical order with blend of Mughal *chatris* and *jalis* (Peck, 2005). The Central Business District of New Delhi, popularly known as Connaught Place was laid out in circular fashion around a Central Park, similar to the Royal Crescent in Bath. The impression of Georgian terrace houses of the Royal Crescent is visible in the façade of Connaught Place, which was to house offices and shops at the ground floor and residences at the first floor. Today, it is strictly in commercial use and the residential floors are converted into cafés, restaurants, shops and offices.

Soon after the inauguration of New Delhi, India became independent in 1947, and the strong architectural imprints of Europe on New Delhi are preserved as memorabilia of profound rulers. By this time, Delhi stood as a confused mosaic of many shapes and forms, from ancient to modern. However, the architectural

Figure 4.71 Viceroy's House (modern-day Rashtrapati Bhavan), an example of European Neo-Classicism and Indo-Saracenic style

Source: Photograph by Harsh Tripathi, 2017

Figure 4.72 Central Secretariat, an example of Neo-Classicism and Indo-Saracenic style

Source: Photograph by Harsh Tripathi, 2017

Figure 4.73 Royal Crescent at Bath (top) and Connaught Place at Delhi (bottom)
Source: Photograph by Harsh Tripathi, 2017

dualities in Delhi, as the capital of Republic of India, are read as the chapters of chronological biography of architectural aspirations of ambitious political powers.

The enthusiasm of Art Deco was unaffected by the sudden upsurge of Neo-Classicism in Delhi, and many buildings of this order occupy the urban landscape of India. However, after making a strong presence, the momentum of Art Deco gradually slowed down in the second half of the twentieth century, which was dominated by modernists, who criticised Art Deco for its whimsical nature (Lang, 2002).

4.11.5 Modernism in India (1930s and 1940s)

Founded on Rationalist ideologies, modernists aspired to break away from the past and use a new architecture that could serve people better. Emphasis was placed on functional efficiency of the building and simplicity of form. As per the Western philosophy, there are two approaches to shape the future: Empiricism and Rationalism (Lang, 2002). Empiricists believe that knowledge should be based on evidence, and Rationalists argue that "truth and beauty can be divined by pure reasoning" (Lang, 2002, p. 3). In the context of architecture, the Empiricists relied on precedents, while Rationalists believed in creating new and original. The International Style was the formative phase of modernism that was practiced between 1920 and 1960 and laid the foundation for majority subsequent works. Going back to the basics, this style used pure geometric forms and resulted in a variety of avant-garde architectural developments that were constrained only by technological reality (Lang, 2002). Also, the ornamentation of buildings, which was common in all parts of the world until the late nineteenth century, was simplified or rather nullified, and the inherent qualities of building materials were utilised for visual attractiveness. This also raised the continuing debate on ornamentation and austerity and opened opportunities for innovations in material and design.

Contrary to the wide range of approaches to modernism in Europe, the focus of Rationalist architecture in India was narrowly set on the works of Bauhaus school (1919–33) and Le Corbusier (Lang, 2002). The ethos of modernism in India was also impacted because of the political divide between architects into nationalists and modernists and therefore 'modernism', being of foreign origin, received only limited appreciation. The majority of modernists' works in India were by foreign architects who were working for wealthy clients, including businessmen and entrepreneurs, who admired modernism (ibid). This also included architects from the Art Deco movement who simplified their designs to come closer to modernism, for example, Walter Sykes George (his modern works are discussed later).

During the 1920s and 1930s, when modern architecture was in a formative phase and the international stage was being set for new architectural and social reforms, the architecture in India was still celebrating the fantasies of Art Deco. Delhi was only partially in terms with Art Deco and was rather embracing Neo-Classicism, chosen as the expression of imperialism. However, the style that Delhi observed in the second half of the nineteenth century grew upon principles of modernism. The 1930s and 1940s were phases of transformation, both political and architectural, in which the continuity of the imperial past overlapped with the ambitious future, the shape of which was only partially known.

4.12 The modern Delhi – post-1947

The end of political ties with the British did not necessarily mean the end of the architectural alliance with European style. Put another way, political discontinuity did not induce architectural discontinuity, unlike in the past. Political aspirations were less imposed on architecture, and instead the discipline was predominantly guided by architectural educators and professionals. Since most of the architects in the country were trained under British teachers at the JJ School, or worked with firms ran by foreign architects, the most influential being Gregson, Batley and King (GBK), the momentum of European style continued for a few decades post-1947 (Lang, 2002). To explain more clearly, there existed three sets of architects and architecture in India at the time of independence: first, the rare but enthusiastic nationalists, who took pride in the past and relied on traditional Indian precedents; second, the admirers of Art Deco, who were active till the 1960s; and third, the modernists. The next sections will discuss in detail the process of architectural evolution of each and the physical impressions these created on Delhi, post-independence. In summary, nationalism would gradually evolve to embrace regional and vernacular architecture; Art Deco would simplify to move closer to modernism; and the modernists, the most influential of all, would dominate the urban scene, and many of them would be renowned for their unique individual style of advancing upon modernism principles.

Lang (2002) categorises modernists in India into two groups, based on the nature of building designed, as well as their sources of architectural inspiration. The 'first-generation modernists' in India were usually trained outside the country and were highly influenced with the works of Frank Lloyd Wright (1867–1959, American) and Walter Gropius (1883–1969, German) (Lang, 2002). Lang (2002) explains that Empiricism of Frank Lloyd Wright gave way to Empiricism of Louis I Kahn (1901–74) and similarly rationalism of Walter Gropius cascaded into Le Corbusier (1887–1965), who were the role models for second-generation modernists. Political freedom for the country also brought along free exchange of information and freedom to choose from among multiple sources of inspiration across the world. And, second-generation modernists were therefore exposed to a much wider array of international architects; Louis I. Kahn (American) and Le Corbusier (Swiss-French) are the most influential for their direct association with India. The involvement of influential international architects in the role of educators, for example, Charles Eames (1907–78) and Ray Eames (1912–88), also widened the horizons for architectural inspirations for Indian architects. Comparing the attitudes of the first- and second-generation architects, Lang (2002) writes that even though the philosophy of the Sir JJ School was to produce modern architecture, with symbolic referents to India that should respond to the social and climatic environment, there is greater sympathy towards these objectives among the later. This is attributed to the dominance of foreign architects in the education and professional sector, as well as to the overall control of the British Empire till 1947. In general, the first-generation modernists were very conservative, and the second-generation architects, many of whom received foreign education after the Second World War, when the Western world was bursting with enlightened thinkers, were much more confident of giving shape to their ideas in a free nation.

4.12.1 *First-generation modernists*

It took a while before political and social unrests could settle post-independence, particularly after the Partition (refer to Chapters 1 and 2 for more details). The socialistic and welfare policies of the government were best accomplished in the austerity of modern architecture. There was an urgent need for administrative and institutional buildings that could house the activities of the expanding democratic government. There was a pressing demand for all types of public and private buildings, like schools, hospitals and housing. The changes to the built environment, caused by upsurge in building activities, were swifter and loftier in Delhi for being the seat of the government and also for being the host to majority population which was dislocated during the Partition. The history of Delhi so far was full of examples of opulent buildings for powerful emperors, and it was probably the first time when public buildings were dominating the scene. The change of political structure, from autocratic to democratic, was directly reflected on the urban fabric of this old, yet modern, city. Given the constraints of resource and the urgent need for public buildings, modernism was not only the style but also the necessity.

Some of the earliest examples of modern buildings in Delhi are the Tuberculosis Association Building (1950–52), the Lodi Colony housing of the late 1940s and the Sujan Singh Park (housing apartments) (Figure 4.74), all by Walter George (Lang, 2002). These buildings are characterised by simplicity of form and layout and plainer façades in brick and stucco. Sujan Singh apartments are also partially Art Deco in the use of the curved entrance façade (Khanna & Parhawk, 2008). While these housing projects catered to households in the higher-middle-income group, there were many other lower- and middle-income group housing projects undertaken by the Central Public Works Department (CPWD) and later by the DDA (Lang, 2002). These projects were mostly located in Lajpat Nagar (Figure 4.75), Malviya Nagar, Rajendra Nagar, Patel Nagar and other newly created neighbourhoods for the refugees (refer to Chapter 3 for more details). The typical manifestation of public-sector housing was very much in the Bauhaus style, as demonstrated by lacking embellishment; emphasis upon functional efficiency of plan; layout of continuous rows of flat roofed houses around open spaces; and verticality broken by horizontal projections, made of thin concrete slabs, placed over windows and porches.

The austerity of modernism was selectively chosen for mundane projects, like the public housing and important government buildings, particularly in the mall, which were still drawing from the Indo Saracenic principles originally used by Lutyens and Baker for New Delhi. Many buildings were designed along similar lines by the CPWD, under the leadership of Ganesh Bhikaji Deolalikar (1890–1979), for example, the Supreme Court (1952), Udyog Bhavan (1957), Krishi Bhavan (c. 1957), Rail Bhavan (1962) and Vigyan Bhavan (1962). The dome of Supreme Court (1952) (Figure 4.76) is in the form of Sanchi Stupa, similar to that of the Viceroy's House, and the Vigyan Bhavan (Figure 4.77) used Tudor arch of Mughals. Modernism was popularly adopted in the design of hotels, colleges, institutions and even factories and power stations (Lang, 2002). A faithful example of the International Style of Gropius is the International (modern-day Oberoi) Hotel (Figure 4.78) at Delhi built in 1958 by Durga Bajpai and Piloo Mody.

Under the leadership of Habib Rahman, an architect with CPWD, the influence of Bauhaus ran into many institutional buildings, which give to central New Delhi the character it has today. Even though Bauhaus masters anticipated it to be a free style, a pattern emerged and characterised Bauhaus with cubist form, flat roofs and façades, minimal ornamentation and shaded windows (Lang, 2002, p. 53).

Figure 4.74 Sujan Singh Park, an example of modernist housing projects by Walter George

Source: Photograph by Harsh Tripathi, 2017

Figure 4.75 Lajpat Nagar housing, an example of a plotted residential development by the public sector

Source: Photograph by Harsh Tripathi, 2017

Figure 4.76 Supreme Court (1952), an example of modern architecture

Source: Photograph by Harsh Tripathi, 2017

Figure 4.77 Vigyan Bhavan (1962), an example of modern architecture

Source: Photograph by Harsh Tripathi, 2017

Figure 4.78 International Hotel (1958), modern-day Oberoi Hotel, an example of International Style

Source: Photograph by Harsh Tripathi, 2017

Figure 4.79 Curzon Road Hostel, Delhi, an example of Bauhaus style

Source: Photograph by Harsh Tripathi, 2017

Figure 4.80 Rabindra Bhavan at Delhi

Source: Photograph by Harsh Tripathi, 2017

Figure 4.81 Kirori Mal College (1956), an example of modern architecture

Source: Photograph by Harsh Tripathi, 2017

Typical examples of Bauhaus in Delhi include the Dak Tar Bhawan (1954), the Auditor and General Collector's Office (1958), the Indraprastha Bhawan (1965), the World Health Organisation Building (1963) and the Curzon Road Hostel (1967) (Figure 4.79) (Lang, 2002, p. 53). Along similar lines, Rahman proposed a box-like structure for Rabindra Bhavan (Figure 4.80) which was strongly discarded by Nehru. Nehru's challenge to Rahman's repetitive Bauhaus design compelled him to introduce new curvilinear flowing form of Rabindra Bhavan. Rahman's subtle use of Indian elements like *chajjas*, *jalis* and overhanging roofs adds to the character of the building (Khanna & Parhawk, 2008)

Rahman's contemporary Cyrus S. H. Jhabvala designed Kirori Mal College (Figure 4.81) in 1956. Lang (2002) doubts that this work has closer resemblance to the modernist architecture of Scandinavian countries rather than with Bauhaus.

Alongside Rationalism developed Empiricist modernism, as seen in institutional buildings – Triveni Kala Sangam (1957) (Figure 4.82), India International Centre (1962), Ford Foundation Building (1969), Unicef building (1988) and India Habitat Centre (1988–93). These all are works of Joseph Allen Stein (1912–2001), who was probably among the few renowned foreign architects employed as educators and practitioners in 1952. His works, inspired from Frank Lloyd Wright's ideas, break away from formal compositions of Rationalists and strongly embrace the 'organic' aesthetic character (Lang, 2002). The use of rustic local building materials and techniques and the strong alliance with nature are two major characteristics in his design. At Triveni Kala Sangam, the open-air theatre is the focal point of the building, and the interaction with nature is enhanced inside the building by cutting off walls wherever possible. The *jalis* on external walls put sunlight on play while also reducing excessive glare. The harmonious juxtaposition of built and unbuilt spaces is achieved very effortlessly. The India International Centre (Figure 4.83) was meant to be a place where various currents of intellectual, political and economic ideas of writers, scholars and academics could interact freely (Khanna & Parhawk, 2008). Design of the building allows free movement between the residential wing and the public facilities wing, which are connected with open passageways and courts as meeting places for formal and informal interactions. This idea is later repeated at the India Habitat Centre built for functionally diverse activities which can be unified through socially interactive spaces, like passages and courts. Subtle use of green ceramic tiles lightens the mass of Lahori brick façade. Most eye-catching is the blue-coloured pergola over the central courtyard, which keeps the space shaded during harsh summers (Figure 4.84). Joseph Stein's expertise in combining local conditions and material to create architecturally profound edifices is highly appreciated by the generations to come. Overall, there is admirable consistency in the architectural theme of his buildings, irrespective of the wide timeframe during which these were constructed.

Indian interpretations of modernist ideas varied significantly, thus adding many new forms and patterns to the built environment. This was the start of eruption of multiple forms, materials and geometries which were not always in harmony with the surroundings. Architectural elements introduced by first-generation modernists were a guide to what followed in the later years.

Figure 4.82 Triveni Kala Sangam (1957), an example of Empirical modern design

Source: Photograph by Harsh Tripathi, 2017

4.12.2 Second generation of modernists (1950–70)

As mentioned earlier, Kahn and Le Corbusier were the main inspiration for the second-generation Empiricists and Rationalists, respectively (Lang, 2002). The most evident influence of Rationalist ideas of Le Corbusier was in the site planning. Le Corbusier used buildings as elements in space rather than space-makers and their use to wall outdoor spaces (Lang, 2002). A good example is the design

Figure 4.83 India International Centre (1962), an example of Empirical modern design

Source: Author, 2016

of Jugal Kishore Chowdhury for the Indian Institute of Technology (IIT) campus (1961) at Delhi. Lang (2002) refers to the courtyard of Oxbridge as a precedent to IIT. Arrays of courtyards were created by parallel placement of three-story academic blocks and the perpendicular positioning of a seven-story block (Lang, 2002). Covered walkways connect the building blocks and form courtyards (ibid). A similar idea runs into the layout of Jawaharlal Nehru University (JNU) campus (c. 1973) at Delhi, which was jointly designed by CPWD and C. P. Kukreja. Taking inspiration from first-generation modernists, the use of these design elements became popular among next generation of modern architects – horizontally expanding building mass with lower plinths; large glass windows in horizontal bands; freestanding staircases; and cantilevered porches, most of which are also used at IIT and JNU (Lang, 2002) (Figure 4.85).

Figure 4.84 India Habitat Centre (1988–93), an example of Empirical modern design

Source: Author, 2016

Figure 4.85 Indian Institute of Technology, Delhi, an example of Rationalist modern architecture

Source: Photograph by Harsh Tripathi, 2017

Figure 4.86 Shri Ram Centre (1969), an example of Rationalist modernism

Source: Photograph by Harsh Tripathi, 2017

Figure 4.87 Akbar Hotel (1969), modern-day Akbar Bhawan, an example of Rationalist modernism

Source: Photograph by Harsh Tripathi, 2017

Among noteworthy examples of buildings constructed during 1960s and 1970s, which draw upon Le Corbusier's works, are the Akbar Hotel (1969) (modern-day Akbar Bhawan) (Figure 4.87) and Shri Ram Centre (1969) by Shiv Nath Prasad (Figure 4.86). The former relies heavily on the Le Corbusier-designed Unite d'habitation (1952) in Marseilles, while the latter is much more original in form. Very faithful to the Rationalist ideology, Shri Ram Centre uses pure geometric forms, each of which is identified with the function they serve. As sculptures of reinforced concrete, these buildings are at times categorised under Brutalist architecture in Delhi. However, Lang (2002) explains that the exposure of building material and construction technique, as observed in buildings from 1970s, was the necessity of austerity more than being the aesthetic of Brutalism. In fact the inspiration for exposing rustic building materials lay in Indian vernacular architecture, and it happened that Brutalism concept in Europe paralleled to what was happening in India for other reasons (Lang, 2002).

Another wing of Rationalists, growing alongside but contrary to the Le Corbusier group, was formed of architects who used innovative building forms to enclose space (Lang, 2002). This was achieved in collaboration with structural engineers, among whom Mahendra Raj gave significant contribution to the building industry (ibid). One among his many noteworthy works is in collaboration with architect Raj Rewal, who designed the Permanent Exhibition Complex (1972) at Pragati Maidan, Delhi (ibid) (Figure 4.88). It is a space frame structure created by the use of an octahedral lattice built on site (Lang, 2002). However, the structure was demolished recently and invited a lot of resistance from the local people and architecture community.

In many extreme cases, innovative building forms were created at the compromise of essential functions of the building, and the continuity of this trend was put to halt.

Figure 4.88 Permanent Exhibition Complex (Pragati Maidan), Delhi, an example of Rationalist modern architecture

Source: Photograph by Saumya Sharma, 2007

The Empiricists of second-generation modernism were more influenced by Louis Kahn even though his architectural association with India was limited to only one project, but worthy enough, the Indian Institute of Management (IIM), Ahmedabad. The juxtaposition of brick buildings and the spaces they create and the play of light and shades created with carefully gauged depths and heights of openings are brilliant elements of design that inspired young architects. An example of Kahn-inspired works in Delhi is the India Statistical Institute (1976) (Figure 4.89) by Anant Raje, who had worked closely with Kahn in IIM project. Humble copy of IIM's arches and circular windows is seen in the Modern School (1978) (Figure 4.90), at Vasant Nagar, Delhi, designed by Sachdev, Eggleston and Associates (ibid).

The years 1970–90 brought along a wave of high-rise commercial buildings, which were modernists and yet visually more appealing. Most of these buildings stood out from other buildings surrounding them for these were either relatively higher or their design configuration was distinct from the neighbours, or for both the reasons (Lang, 2002, p. 93). Examples of point blocks (or towers) are the Delhi Development Authority (Figure 4.91) headquarters designed by Habib Rehman for the CPWD; the State Trading Corporation Building (1989) by Raj Rewal (Figure 4.92). Among

Figure 4.89 Indian Statistical Institute (1976), an example of Empiricist modern architecture

Source: Photograph by Harsh Tripathi, 2017

Figure 4.90 Modern School (1978) at Vasant Nagar, an example of Empiricist modern architecture

Source: Photograph by Harsh Tripathi, 2017

Figure 4.91 Delhi Development Authority headquarters, an example of point blocks (or towers)

Source: Photograph by Harsh Tripathi, 2017

examples of distinct design configuration are the Delhi Civic Centre (1965–83) (Figure 4.93) by Kuldip Singh, which takes inspiration from the observatory of Jantar-Mantar near which it is located; and the Life Insurance Corporation (LIC) (1989) (Figure 4.94) by Charles Correa that uses combinations of brick, glass and metal, all distinctly appreciated.

Charles Correa (1930–2015) was among the Rationalist modernists who could create individual identity in his work, which gained appreciation in India and abroad. He embraced the Gandhian philosophy of simple living and amalgamated it beautifully into modern artistic ideas. Together, these concepts are expressed in

Figure 4.92 State Trading Corporation Building (1989), an example of point blocks (or towers)

Source: Photograph by Harsh Tripathi, 2017

the use of simple forms and frugal means for achieving architectural ends (Lang, 2002). Although his works in Ahmedabad are a more sincere adoption of these philosophies, the LIC (1989) building at Connaught Place, Delhi, demonstrates a deviation, and thus his versatility (Khanna & Parhawk, 2008). Its modern façade in sandstone, metal and glass façade is in sharp contrast to the existing circular precinct, though similar to other high-rise buildings that emerged in the 1980s. On top of the structure runs a pergola, and this became the identity of the

Figure 4.93 Delhi Civic Centre (1965–83), near Jantar Mantar, an example of point blocks (or towers)

Source: Photograph by Harsh Tripathi, 2017

building. This is not the best portrayal of Correa's work, and rather his mention in the context of Delhi is more often in reference to Tara Housing (Figure 4.96). He was one among the rare, home-bred, protagonists of Indian architecture community to whom the industry looks upon for inspiration.

Increased participation of Indian architects among second-generation modernists was building up the confidence of young Indian architects in home-bred

Figure 4.94 Life Insurance Corporation building Jeevan Bharti (1986), an example of Rationalist modernism

Source: Photograph by Harsh Tripathi, 2017

architecture. While the osmosis of international ideas in India continued to happen, its appropriation and contextualisation by Indian architects increased the admissibility. In addition to that, Indian architects were quick enough to build a reputation for their individual work, both nationally and internationally. International acknowledgement of Indian architects was even more necessary for their image building in the home country. The shift from foreign ideals to local role models was an important achievement by second-generation architects.

An interesting observation into the nature of buildings produced between 1950–90 tells that majority were for the public sector, mostly institutional clients, and only a few private projects were undertaken. This was the developmental phase of democracy and welfare society in the country, and therefore government was carefully regulating private sector activities. In 1991, the economic liberalisation of the country opened up many opportunities for the private sector, and this was directly visible in increased building activity by and for the private clients.

4.12.3 Post-modernism and neo-traditionalism in Delhi

Abstract-ism from the past, as the core of post-modernism, has been observed in the architecture of Delhi since the beginning, whether it was the incorporation of abstract Hindu elements in the Islamic architecture, or the use of Hindu and Saracenic features in British monuments, like the Viceroy's House. Among modern architects, Habib Rahman often used abstracts from Islamic architecture, for example in the Mazar of Maulana Azad (1960) (Figure 4.95), Rabindra Bhavan (1961) (Figure 4.80) and the Mazar of President Fakhruddin Ali Ahmed (1975)

Figure 4.95 Mazar of Maulana Azad (1960) (top) and Mazar of President Fakhruddin Ali Ahmed (1975) (bottom), examples of post-modern architecture

Source: Photograph by Harsh Tripathi, 2017

(Figure 4.95) (Lang, 2002), most of which was designed even before 'post-modernism' gained foot in 1966 (ibid).

However, 'neo-traditional' was believed to be a better response to the idea of incorporating the past (ibid). Three major events in late 1960s suggested that referring to the past architectural heritage within the cultural context of the natives

Figure 4.96 Tara Housing (1975–78) by Charles Correa, an example of neo-traditional Indian architecture

Source: Photograph by Harsh Tripathi, 2017

might be the way forward for Indian architecture (Lang, 2002). First, there was a growing realisation across countries that modernism has failed in achieving the objectives in its manifesto and that generic building pattern and site layouts shall be accustomed to the local climate and conditions (ibid). Second, environmental behaviour studies were beginning to contribute to architectural education and narrow approach of modernism towards building functions was criticised (ibid). Adaptation to local climate and environment was giving further drive towards traditional architecture (ibid). Third, after the exhibition of 'Architecture without Architects' at New York in 1964, a further boost to traditionalist ideology was provided due to the popularity of Hasan Fathy's work on Egypt (ibid). There were two distinct approaches adopted to learn from the past: one who researched the patterns of settlements, building materials and technologies used in the past, defined as vernacularists; and second, who revived the canonical texts such as *Shilpa Shastras* (ibid). The latter group was passively contributing throughout this time, particularly in southern India. There is limited example of the latter type in Delhi, whereas the impact of the former was strongly felt.

Settlement patterns and social symbiosis, light and shade pattern, construction techniques and building materials are the core areas of study of vernacular architecture. These influence housing developments in particular, of which many examples are observed in Delhi. Raj Rewal's design of French Embassy Staff Quarters (1967–69); Asiad Village (1980–82) for Asian games and Sheikh Sarai (1980) are typical examples, which take inspiration from traditional housing settlements of Jaipur and Jaisalmer, Rajasthan. Other important examples are the Tara Housing (1975–78) (Figure 4.96) by Charles Correa and Yamuna Housing Society (1973–80) by the Design Group. Common element across these designs was the use of internal street network as the main access to the house. Quasi-public nature of narrow residential lanes of old Indian towns could not be replicated even through the imitation of physical design. In most cases, the entry from the parking behind was used as the main access, and this defeated the purpose of the front street as a place for social interaction.

The definition of post-modernism in India expanded to include physical design solutions to the problems of the modern society, while taking inspiration from the past. Lang (2002) writes that as post-modernism matured between 1975 and 1995, imitation from the West decreased, and a number of new, contextual approaches to design emerged. Thus, India was finally experiencing architectural liberalisation.

4.12.4 Liberal Delhi – post-1991

Economic liberalisation in India was contemporaneous to its architectural liberalisation from non-contextual adaptation of foreign architectural styles. When economic liberalisation opened opportunities for the private sector to participate and compete in the open market, a new set of consumers was made available. In the world of architecture, Indian architects had already earned significant reputation for their individual style and were familiar with the wide set of tried and tested methods from the past. Therefore, Delhi, since liberalisation, offers conducive market where a variety of clients are meeting versatile group of architects and a symbiotic relationship is harnessing. The street stretches of Huaz Khas village and DLF City (Figure 4.98) are good examples of aspirations of private clients and architect's response to that. The outcome, in the form of commercial and residential structures, is significantly different from the one achieved by public-sector clients a few decades ago.

Figure 4.97 Shopping complex at Vasant Kunj (bottom), an example of public-sector development; Ambiance Mall (top), Gurugram, an example of private development

Source: Photograph by Harsh Tripathi, 2017; author, 2017

Figure 4.98 Plotted development by private developer at DLF City, Gurgaon
Source: Photograph by Harsh Tripathi, 2017; author, 2017

In the history of Delhi's built environment, buildings have been the face of the strongest stakeholder in the society. Starting from the Sultans to the Mughals and the British, the built environment of Delhi has always responded to its administrators. Post-independence Delhi became the seat of political power, and its built environment was largely dominated by public-sector buildings. Even though liberalisation declared the market to be the strongest player, the nature of private buildings in Delhi is doubted as true reflections of its owners. Put another way, development control regulations at Delhi might not be the best example of effective city development strategy, thus at times bringing down the ambitious designs proposed by private clients. Dominant role of government in regulating the built environment of Delhi is indirectly stated in the fact that even though there is a strong demand for space from the private sector who also have the financial capacity to build as per their aspirations and requirements, the highest tower in Delhi (102 metres and 28 stories) is the Civic Centre for Municipal Corporation of Delhi (2010) (Figure 4.99). This brings back the discussion on non-design determinants of design, which, in the present date, are probably the institutions of land and urban planning, building technology and, above all, the economy.

Post-liberalisation, the development authority at Delhi has strictly guarded the city, particularly the New Delhi area, against the growing demand for space in the region. This has caused a sudden increase in building activities in the neighbouring cities of Gurgaon, Noida, Faridabad and Ghaziabad. While these cities were less prepared to take the load, they are slowly responding to the changing needs. However, the slow improvement in the situation after 26 years of liberalisation demands inquiry on the approach adopted towards development in and around Delhi (Rao, 2016). There is a significant disparity in the quality of the built environment within the city, and addressing these concerns shall require wider collaboration between professions (Lang, 2002). As mentioned earlier in the chapter, there are many non-design determinants of design and therefore the

Figure 4.99 Municipal Corporation of Delhi Civic Centre (2010), an example of point blocks (or towers)

Source: Photograph by Harsh Tripathi, 2017

solutions to challenges posed to the architects extend beyond the scope of their profession, as witnessed at Delhi. Collaborative response of polity, society and economy together shall better the quality of life at Delhi in the future.

4.13 Conclusion

From its first impression, Delhi looks chaotic. When seen through a bigger lens, the minutest reasons are made visible, and the city is then appreciated for hosting the most complex makeovers. There were several stages of political and social transformation through which the city underwent, before it achieved a harmoniously unified state of fusion, if at all. The architecture of this city is the continuum from the past which overlaps with the present to create aspirations for the future. Recapturing Ray's (1964) discussion, the complexity of society is reflected in the multiplicity of building typologies. Although her observations were in the context of pre-historic civilisations, similar relationships can be drawn for later societies, albeit with caution. Here complexity will mean the cultural and religious diversity in the society; attitude of the government and its relationship with the governed; and hierarchies and stratifications based on social status, caste, religion, region, dynasty and so on. Considering five social transformations corresponding to five major political changes in Delhi, a summary of building typologies under each is presented in Table 4.1.

The table is suggestive and does not contain a full range of building typologies built during different eras. Rather, it is a summary prepared on the basis of discussions in this chapter. With some reservations, it can be said that the society was becoming complex with time due to the additional religion and culture that was being introduced at each stage. Every time new societies and buildings were founded, a new band was introduced to the existing spectrum, thus widening its hues and adding to the level of 'complexity'. Put another way, the introduction of each new layer added to the existing layers of people, culture, religion and buildings, added to complexity of the built environment as well as the society. Many layers of past civilisations and their buildings coexist at Delhi. This is a very simplified explanation of various strands of social and urban fabric of contemporary Delhi.

Architectural history of Delhi discusses the additional, non-functional, role of buildings, which at times superseded their functional fulfilments. This asserts multiple definitions of building as follows: Iconography of religious ideologies; edifice of political aspirations; instrument of political control; demonstrator of wealth and opulence of the patrons; communicator of aesthetic taste and architectural maturity of its builders; statement of cultural supremacy; written records of social and political environment as well as architecture style and construction techniques; experiments of structural engineers; register of foreign visitors and their gifts including their architectural style; source of pride and self-esteem; stimulator of nationalist movement; and so on. Given the multiple roles of a building, there are many social, political, religious and other non-design determinants which inform design and architecture, as discussed throughout in this chapter. In the contemporary society, buildings are often equated with the economic status of its owners and occupiers, thus raising competition to be unique and identifiable, at times to the extent of being non-contextual to the social and natural environmental where it exists.

Architecture is to be redefined as a language and buildings as its vocabulary, which communicates the needs, desires and aspirations of the society at any given point in time. In the dynamic design process, new vocabulary is added every time a building is created. These put together form the architectural history of a city like Delhi.

Table 4.1 Temporal changes in the level of complexity of the society and building typologies

Complexity of society ∝ building typologies				
Least complex				*Most complex*
Hindu Rule *(Till 1206)*	*Sultanate Rule* *(1206–1526)*	*Mughal Empire* *(1526–1858)*	*British Raj* *(1858–1947)*	*Democratic Republic* *(1947 onwards)*
1. Temples 2. Stupas 3. Monasteries 4. Shrines 5. Military forts 6. Fortified palaces 7. Fortified cities 8. Stepwells (or Baoli)	9. Mosques 10. Tombs, cenotaphs and mausoleums 11. Shrines (or dargah or *mazar*) 12. Minarets 13. Military camps and camp cities 14. Fortified palaces 15. Fortified cities 16. Town gates 17. Schools of theology (*madrasa*) 18. Stepwells, dams and reservoirs	19. Mosques 20. Tombs, cenotaphs and mausoleums 21. Shrines (or dargah or *mazar*) 22. Minarets 23. Military camps 24. Camp cities; fortified cities 25. Fortified palaces 26. Town gates 27. Hall of Public Audience 28. Hall of Private Audience	38. Churches 39. Military cantonment 40. Civil line 41. Viceroy's House 42. Secretariat 43. Residential developments for government officers and staff 44. Public sector offices 45. Town hall 46. Museum 47. Library 48. Ball room 49. Railway stations and tram routes 50. Memorials 51. Clock tower 52. Coffee houses 53. Hotel	54. Religious buildings 55. Military cantonment 56. Rashtrapathi Bhavan 57. Secretariat 58. Residential developments for government officers and staff 59. Public sector offices 60. Embassies 61. Private sector offices 62. Private resorts 63. Clubs 64. Facilities inside gated communities 65. Public transport 66. Memorials and tombs of public figures 67. Housing by public sector 68. Slums and informal housing 69. Resettlement colonies for the refugees 70. Housing by private sector, usually gated communities; memorials (including *smarak*)

29. *Hammam* (or bath) to host formal confidential discussions 30. Hunting resorts 31. Leisure pavilions 32. Gardens 33. Ornamental water bodies 34. Schools of theology (*madrasa*) 35. Stepwells, canals, wells, dams and reservoirs 36. Bazaar or market places 37. *Serai* (or rest houses for travellers)	71. Central Business District 72. Convention and exhibition centres 73. Hotels 74. Museums 75. Movie theatres 76. Public gardens 77. Sports stadia

Source: Author

Notes

1 Defined later in the chapter as an ornamentation style that uses interlaced vegetal forms and interlocked geometric shapes and patterns.
2 John Murray (1911).
3 Fourth city of Delhi founded in 1325 (Blake, 1991).
4 Fifth city of Delhi founded in 1354 (Blake, 1991).
5 Merklinger (2005).
6 Other rulers of these dynasties were buried outside Delhi, and the influence of provincial architecture was probably dominant on their tomb designs.

Bibliography

Abdullahin, Y., & Embi, M. R. (2015, March). Evolution of Abstract Vegetal Ornaments in Islamic Architecture. *International Journal of Architectural Research*, 9(1), 31–49.

Abdullahin, Y., & Embi, M. R. (2013). Evolution of Islamic Geometric Patterns. *Frontiers of Architectural Research*, 2, 243–251.

Ackerman, J. S., Gowans, A. & Collins, P. (2017, July 13). *Architecture (Mimetic Ornament)*. Retrieved from Encyclopedia Britannica: www.britannica.com/topic/architecture/Mimetic-ornament

Alam, M. (1997). State Building Under the Mughals: Religion, Culture and Politics. *Cahiers d'Asie Centrale*, 105–128. Retrieved July 29, 2017, from https://asiecentrale.revues.org/478#authors

Asher, C. B. (1992). *The New Cambridge History of India: Architecture of Mughal India*. Cambridge: Cambridge University Press.

Beglar, J. D., Carlleyle, A. C., & Cunningham, A. (1874). *Report for the Year 1871–72, Delhi and Agra*. Calcutta: Archaeological Survey of India.

Blake, S. P. (1991). *Shahjahanabad: The Soverign City in Mughal India, 1639–1739*. Cambridge: Cambridge University Press.

Brown, P. (1956). *Indian Architecture (Islamic Period)*. Bombay: D. B. Taraporevala Sons & Co. Pvt. Ltd.

Burton-Page, J. (2008). *Indian Islamic Architecture: Forms and Typologies, Sites and Monuments* (G. Michell, Ed.). Leiden: Brill.

Crane, H. (1993). Notes on Saldjūq Architectural Patronage in Thirteenth Century Anatolia. *Journal of the Economic and Social History of the Orient*, 36(1), 1–57.

Department of Archaeology and Museums. (2004, January 30). *Tomb of Shah Rukn-e-Alam*. Retrieved July 20, 2017, from UNESCO World Heritage Centre: http://whc.unesco.org/en/tentativelists/1884/

Department of Islamic Art. (2001, October). *Vegetal Patterns in Islamic Art*. Retrieved July 29, 2017, from Heilbrunn Timeline of Art History: www.metmuseum.org/toah/hd/vege/hd_vege.htm

Golcha Cinema. (2007). *Home*. Retrieved August 1, 2017, from Golcha Cinema: Since 1954: www.golchacinema.com/

Grover, S. (1996). *Islamic Architecture in India*. New Delhi: Galgotia Publishing Company.

Gye, D. H. (1988). Arches and Domes in Iranian Islamic Buildings: An Engineer's Perspective. *British Institute of Persian Studies*, 26, 129–144.

The Imperial India. (2017). *History of the Imperial*. Retrieved August 1, 2017, from The Imperial: www.theimperialindia.com/imperial_history/

Khan, S. (2016). *History of Islamic Architecture: Delhi Sultanate, Mughal and Provincial Period*. Delhi: CBS Publishers and Distributors.

Khanna, R., & Parhawk, M. (2008). *The Modern Architecture of New Delhi (1928–2007)*. Noida: Random House India.

Kleiss, W. (2011, October 28). *Construction and Materials Techniques in Persian Architecture*. Retrieved June 10, 2017, from Encyclopaedia Iranica: www.iranicaonline.org/articles/construction-materials-and-techniques-in-persian-architecture

Lang, J. (2002). *A Concise History of Modern Architecture in India*. New Delhi: Permanent Black.

Lang, J., Desai, M. & Desai, M. (1997). *Architecture and Independence: The Search for Identity India 1880 to 1980*. New Delhi: Oxford University Press.

Luniya, B. N. (1978). *Life and Culture in Medival India*. Indore: Kamal Prakashan.

Merklinger, E. S. (2005). *Sultanate Architecture of Pre-Mughal India*. New Delhi: Munshiram Manoharlal Publishers Pvt. Ltd.

Michell, G., & Pasricha, A. (2011). *Mughal Architecture & Gardens*. Suffolk: Antique Collectors' Club.

Murray, John. (1911). *A Handbook for Travellers in India, Burma, and Ceylon*. London: John Murray.

Othmann, Z., Aird, R. & Buys, L. (2015). Privacy, Modesty, Hospitality, and the Design of Muslim Homes: A Literature Review. *Frontiers of Architectural Research*, 4, 12–23.

Oxford Dictionary of Islam. (n.d.). *Tawhid* (J. L. Esposito, Ed.). Retrieved July 13, 2017, from Oxford Islamic Studies Online: www.oxfordislamicstudies.com/article/opr/t125/e2356?_hi=0&_pos=2

Patel, A. (2004). Toward Alternative Receptions of Ghurid Architecture in North India (Late Twelfth-Early Thirtheenth Century CE). *Archives of Asian Art*, 54, 35–61.

Peck, L. (2005). *Delhi: A Thousand Years of Building*. New Delhi: Thomson Press.

Prater, R. H. (2017, May). *The Neo-Georgian Architecture of Sir Edwin Landseer Lutyens (1869–1944)*. Retrieved July 29, 2017, from Georgia Tech Library: https://smartech.gatech.edu/handle/1853/58268

Rao, J. (2016). Questioning the Approach of Indian Cities Towards Development. In P. Tiwari (Ed.), *The Towers of New Capital: Mega Townships in India* (pp. 111–119). London: Palgrave Macmillan.

Ray, A. (1964). *Villages, Towns and Secular Buildings in Ancient India: c. 150 B.C.–c. 350 A.D.* Calcutta: K.L. Mukhopadhyay.

Shah, J. (2008). *Contemporary Indian Architecture*. New Delhi: Roli Books.

Sharma, Y. D. (2001). *Delhi and Its Neighbourhood*. New Delhi: Archaeological Survey of India.

Smith, K. (2012). *Introducing Architectural Theory: Debating a Discipline*. London and New York: Taylor & Francis.

Tabbaa, Y. (2001). *The Transformation of Islamic Art During the Sunni Revival*. Seattle and London: University of Washington Press.

Welch, A. (1993). Architectural Patronage and the Past: The Tughluq Sultans of India. In M. Sevcenko (Ed.), *Muqarnas Volume X: An Annual on Islamic Art and Architecture* (pp. 311–322). Leiden: E.J. Brill.

5 The intersection of surrealism, welfarism and consumerism

5.1 Introduction

The present urban form, built structures and the image that they present of Delhi and its surrounds are an outcome of many centuries of continual human interventions into the natural environment. The social, cultural, political and economic ideologies of incoming rulers and residents of Delhi have been at times at odds with the existing ones, leading to a prolonged period of experimentation and innovation. The change in the society and hence the built environment, however, seems rather continual than discrete or disruptive, and the elements of built environment from the past have generally been carried into the future with elegance. The visible evidences of millennium of human activity that dot the landscape of Delhi and its surrounds are at times at ease and at other times in conflict with each other. Today, the space seems continuous when the hustle and bustle of Shahjahanabad and Daryaganj transitions into the calm milieu of New Delhi, however, with a bit of unease at the edges. Of course, one experiences these rough edges as two spaces, seemingly from different periods, meet each other. The most noticeable are how the population density changes suddenly, how the narrow roads of Shahjahanabad give way to wide boulevards of imperial New Delhi, how the human-pulled rickshaws give way to cars, how the skyline of colonial bungalows of British period with large gardens and boundary walls gets dotted with tombs of Lodhi, Saiyyads, Humayun, Safdarjung and others in fine Islamic architectural styles of Sultanate and Mughal periods. The buildings of the past seven decades of republic in India stand side by side to the buildings that were constructed many centuries ago – high-rise modern township of Gurugram near Mehrauli (location of Lal Kot, Qila Rai Pithora and Qutub complex), Delhi's first city; Firoz Shah Kotla Cricket Stadium next to the structures from the city of Firozabad, Delhi's fifth city; the symbol of modern India, Pragati Maidan (pavilions for housing trade fairs and conventions) constructed in the 1970s within the boundaries of Dinpanah opposite Purana Qila, Delhi's sixth city; housing colonies of resettled migrants after the Partition in 1947 in Model Town near the old British Civil Lines to the east of Shahjahanabad, Delhi's seventh city – present an epitome of assimilation of spaces to present a vibrant and enviable living built environment. What is important here is to note that these historical spaces have found new uses in Delhi. For example, Lodhi Gardens in Delhi, which housed Lodhi tombs, is now an important space for public gathering and socialisation. The villages such as Nizammudin near Humayun's Tomb and Mehrauli near Qutub complex in the heart of modern Delhi retain the characteristics such as narrow alleyways and non-conforming buildings of earlier periods when motorised transport was unknown and communities were agrarian.

The objective of this chapter is not to provide a concluding summary on built environments in Delhi and its surrounds, based on discussions in earlier chapters, as it will be too ambitious. Rather, we extract key narratives that emerge from earlier discussions that illustrate that a narrow perspective of one discipline such as planning or architecture bounded by limited approaches and time frame for enquiry to observe a phenomenon is insufficient to explain the continuum of the built environment that evolves in time and space.

The periods in the history of the built environment of Delhi and its surrounds could philosophically be described as the periods of 'surrealism', 'welfarism' and 'consumerism'. Breton et al. (1969) argue that

> surrealism is based on the belief in the superior reality of certain forms of previously neglected associations, in the omnipotence of dream, in the disinterested play of thought. It tends to ruin once and for all other psychic mechanisms and to substitute itself for them in solving all the principal problems of life.

Welfarism is based on the premise that actions, policies, rules should be evaluated on the basis on their consequences – intended or unintended. The objective of the approach is to look at the human welfare. Sen (1979) defines welfarism as "the judgment of the relative goodness of alternative states of affairs must be based exclusively on, and taken as an increasing function of, the respective collections of individual utilities in these states". In the words of sociologists Wright and Rogers (2015),

> Consumerism is the belief that personal wellbeing and happiness depends to a very large extent on the level of personal consumption, particularly on the purchase of material goods. The idea is not simply that wellbeing depends upon a standard of living above some threshold, but that at the centre of happiness is consumption and material possessions. A consumerist society is one in which people devote a great deal of time, energy, resources and thought to "consuming". The general view of life in a consumerist society is consumption is good, and more consumption is even better.

These three philosophical narratives could apply to the periods of the built environment development in Delhi and its surrounds in an interesting way. The pre-independence built environment period, which lasted more than a millennium, rulers – early Hindu, Sultanate, Mughal and British – built the city to satisfy their personal vision, self-interest and indulgences, ambitions, thereby determining the style of buildings and kinds of cities that were built. They borrowed from the past and contemporary styles that were prevalent in India as well as elsewhere, adopted and improvised it and built what they thought would best represent their ideologies, power and wealth. This period can loosely be referred to as 'surrealism', more from the interpretation of intent of rulers' point of view rather than the art and architectural style that has been termed as 'surrealist'. The buildings in India that were built during pre-independence time were nowhere comparable to the 'liberating architecture' styles of surrealism of 1930s that swept Paris and other parts of the world.

Post-independence, democratic government had to balance the development objectives with welfare objectives. It had to house millions who became new residents of capital of modern India. Challenge, of course, was that resources were limited. Though government could procure land using compulsory acquisition instruments, a legacy of British Empire, the resources required to build on them were few. Democracy also meant that one could not displease masses in fulfilling the ambitions by acquiring more and more land by force or threat. The leaders of modern India had the experience of harsh consequences of market economy when the British East India Company was at the helm of affairs in the country, and they did not want key resources of the economy to be in the hands of private sector or foreign investors. They thought that for equitable democratic society, it is better if government takes on the role of planner and developer as well. A number of institutions in the public sector were set up, but over time their efficiency and efficacy became questionable. The growth of private sector was stifled through various regulations (discussed in earlier chapters) that were imposed more so in the capital city of Delhi than other parts of the country. Welfarism in India, instead of helping the poor, marginalised them further as the supply lagged demand, and informal settlement was considered intolerable in policies and was considered better to be removed. Development of Delhi's built environment took a back seat.

The third narrative of the built environment in Delhi and its surrounds is associated with consumerism, the period after 1990s. A movement towards consumerism had its silent beginning in 1970s and 1980s when Delhi imposed restrictions on private development. This crowded out private sector that then initiated development on the fringes of Delhi, on land which technically was part of other states that had fewer restrictions (According to the Constitution of India land is a state subject in India). Private sector-led development, however, became a major movement after 1990 when India liberalised its economy and more so in 2000s when the international investors' interests grew in India. This liberalisation brought capital and people on the fringes of Delhi looking for investment and a place to live. Delhi and its surrounds became the hub of unprecedented development activity. While Delhi was still reeling under the regulatory shackles, development on fringes boomed, which created an image of built form that can identify itself with 'consumerism' at its core.

In this final chapter of the book, we continue with discussion on the framework that we introduced in Chapter 1 with reengineered focus to populate the elements of framework with observed trends over time. The section will also validate our approach to involve principles of social theory in understanding the built environment. This will be followed by a discussion on urban form of Delhi and its surrounds after the independence. The discussion also summarises key trends in structures and products of built space during that period. Section 5.4 discusses the pattern of urban form and buildings that are being built in the twenty-first century. The section also reflects on the sustainability of current development practices. Finally, the concluding section presents thoughts on how Delhi has embraced disruptions in time and space.

5.2 Dreams – surrealism

The buildings and cities of thousand years that preceded independence reflect the opulence in material, design and craftsmanship, which can only be explained as 'surrealistic' when compared with contemporary development. When land was abundant and wealth was concentrated in the hands of rulers, it was possible to execute whimsical imaginations on ground. The competition to outdo what has been built in the past or is being built elsewhere further extenuated the grandeur of these schemes. The buildings were identities of those who commissioned them about whom people would revere and travellers would write about. These were also iconic buildings that represented the arrival and establishment of a new religion, which had to be declared aloud. When resources were abundant, austerity took a back seat.

For many of these historical buildings, the functional use became obsolete over time. However, these have organically regenerated in new uses. For example, the privately enclosed areas of Humayun's Tomb and Safdarjung Tomb, which had opulent gardens, have gradually transformed into popular public spaces. The Hauz Khas reservoir and associated buildings, which once were the summer resort for the rulers, are now vibrant urban spaces that are dotted with designer show rooms, art galleries and restaurants.

This period represents the era of design innovations. Complex geometric elements of design and gravity-defying ornamentation became part of Mughal buildings. The use of foliate marble arches in Diwan-i-Khas, Red Fort and 6 and 8 point *jalis* of Humayun's Tomb are some of the examples of these innovative designs. The innovative use of trabeated structural system for arches and domes in Qutub complex, as a way to build quickly with local artisans, and later evolution into a mature arcuated system as in Alai Darwaza within the same complex built few years later, are some other examples of maturity in design. The elaborate water channels inside palatial complexes of the Red Fort are not only ornamental but are also efficient cooling systems. The innovation in design is also seen in the fusion of various architectural styles in buildings of these periods. For example, Bangla roof, originally an architecture style of Bengal Province, became a popular element of Mughal architecture, and was used over the marble throne of Shah Jahan. One of the most important buildings constructed during the British imperial period, the Viceroy's House, takes inspiration from Sanchi Stupa of Buddhist architecture of the third century BC. There are many such innovations that became characteristic of the period and have been discussed in Chapter 4.

The innovation in design was complemented by the introduction of new materials as they became available or prosperity allowed the patrons to procure them. The densely ornamented stone edifices of Hindu temples are direct descendants of wooden proto types that were used in the past. The geometric ornamentation, which is a key characteristic of Islamic architecture, is believed to have descended from brick construction in Central Asia and Iran. The technique of *pietra dura* in marble was another major innovation in material and design employed during Shah Jahan's period (Figures 4.50 and 4.51). Even during the decline of Mughal power and wealth, there was continuity in the use of intricate ornamentation albeit in stucco as visible in the interiors of Safdarjung Tomb. British imperial design was a bold depiction of Greco-Roman features of European styles which slowly started interacting with Indian Saracenic native architecture.

Settlement pattern and layout of cities, which looks organic today, were often a futuristic manifestation of cosmological tenets of the society. When land was abundant, it was possible to give shape to these thoughts. The questions on their viability and appropriateness became secondary as the rulers were autocrats and the opportunities to conduct new experiments were unconstrained. Each ruler in Delhi created a new city of his own desire as if the old were inadequate to showcase his alter ego. This was even true for the British imperial administration despite their long history of representative monarchy at home. Often these new cities were not geographically contiguous with the old, but that did not deter them from experimentation. The best representation of political aspirations and cultural supremacy communicated through design is seen in the case of Shahjahanbad, and later at New Delhi. Shahjahanabad is the realisation of Shah Jahan's aspirations of an imperial capital city. The city thus contains an extensive *serai*; the moonlit bazaar with canal-laced streetscape – Chandni Chowk, one of the largest markets in Asia in the sixteenth and seventeenth centuries, culminating at the royal palace (Red Fort) and terminating at Fatehpuri Masjid. This was the perfect backdrop for royal demonstration of the most powerful emperor in the world. The benchmark set by Shahjahanabad had to be met by the new capital of the British. While Shahjahanabad was still relying on the Persian ancestors, New Delhi took inspiration from recently developed cities of the West, particularly Washington DC.

Table 5.1 summarises the institutions that shaped the built environment during pre-independence period.

Table 5.1 Institutional pyramid for the built environment in Delhi

	Hindu	*Sultanate*	*Mughal*	*British Raj*
Built environment	Mimic ornamentation from wood to stone	Introduction of arcuated system in stone	Splendid use of marble, *pietra dura* ornamentation	Gothic, and classical European styles
Processes	Private prayers required small but numerous praying spaces Modest homes of rulers and nobility Asymmetrical patterns, more organic	Congregational prayers and activities required large gathering spaces Funerary spaces Places for theological education Places for foreign travellers and merchants Grand forts and palaces, symmetric design	Minarets Military camps and camp cities Detailed layout for palaces separating administrative and personal use Recreational spaces for exclusive use Commercial spaces like bazaars and guest houses	Military cantonment and Civil Lines Churches Administrative buildings Public transport and infrastructure

	Hindu	*Sultanate*	*Mughal*	*British Raj*
Agencies	Autocratic political structure	Autocratic political structure Religious norms for ruling	Autocratic political structure Some separation of administration from religion	Colonial administration
Markets	Limited private property rights Layers of use rights granted by the ruler	Centralisation of private property rights Layers of use rights granted by the ruler	Layers of use rights granted by the emperor Rich material and highly skilled craftsmanship	Private property rights Layers of use rights Crystallisation of land revenue system and formalisation of land records
Environment	Autocratic	Autocratic	Autocratic	Colonial

5.3 Welfarism

The austerity of modernism movement of architecture that followed independence was complementary to the welfare policies of newly formed Republic of India. The democracy further put a check on what could be built, and at what cost. The immigration after Partition further strained the resources that were available for developmental activities. Priorities shifted from constructing iconic buildings to build for the masses who desperately needed houses, schools, hospitals and other such basic infrastructure. The government was obliged to meet the social needs though it used similar instruments, as the predecessors, to acquire private land. The objectives of public provisioning of housing and basic amenities remained elusive due to shortage of resources and systemic inefficiencies, resulting in mushrooming of informal housing settlements. The private sector was strictly regulated in its role in the development of the built environment.

Due to this, the role of institutional clients determined the nature of buildings, both in its style and in its functions. Though new uses came in, like art galleries, museums, theatres, convention and exhibition centres, educational institutions and so on, there is an apparent monotonicity and tiredness in the physical outlook and fabric of buildings. Brick and mortar were the medium of expressions, and an overall shift from imperialistic to welfaristic functions and forms was made.

5.4 The built spaces of 'consumerism': malls, metro and *Mal à propos*

The modern-day 'surrounds' of Delhi represent the consumerist side of a city that was planned with a heavy hand after independence. These surrounds were rural hinterlands of the bygone era that shockingly transformed after 1990. This section discusses the transformation with Gurugram (formerly, Gurgaon) as an example. A similar pattern could be observed in other fringe cities such as Noida, Faridabad and Ghaziabad.

The population of Gurugram (to the southwest of Delhi, beyond Mehrauli and Lal Kot) increased from 135,884 in 1991 to 1,514,000 in 2011, a growth of more than tenfold (Mehtani, 2012). The transformation of rural Gurugram on the fringes began in 1980s when, with the onset of liberalised policies, automobile industries started to function there. The car manufacturer Maruti (a joint venture between the government of India and Suzuki Japan) located its plant in Gugugram. This was followed by setting up of ancillary industries and other automobile industries. Major developers who had acquired huge tracts of land in villages in Gurugram, which were on fringes of Delhi, started lobbying with the state government to convert these lands into residential and commercial uses. In 1981, Haryana Urban Development Authority (HUDA) under Haryana Urban Development Act was set up and a private developer, DLF, got the first license issued by HUDA to develop 39.34 acres of land as a residential complex and sell properties to private individuals in Gurugram (Mehtani, 2012). Initial activities were around the old Gurugram town that emerged as the automobile cluster, but later the axis of development shifted to the fringes of Delhi. Private developers determined the direction of growth and development (ibid).

The 1990s brought a shift towards services sector industries. The first to locate was the business processing office (BPO) or knowledge processing office (KPO) of General Electric (GE). GE was looking for a location to set up its India operations, and DLF convinced it to begin operations in Gurugram. This was followed by a number of multinational corporations such as Microsoft, Ericsson, American Express, IBM, Bank of America and American Airlines during the decade that followed, making Gurugram the call centre capital of India (Mehtani, 2012). The unceasing construction activities continued, well assisted by HUDA. The number of settlements more than doubled during 1991–2011, and the DLF City became Asia's biggest private township (ibid). The earlier unfamiliar built forms emerged in Gurugram, catering to the large immigrant population and modern industries that located there (Figures 5.1, 5.2 and 5.3). These new developments were delinked from the local context and were at odds with the development that had followed since 1980s when HUDA sold subdivision residential plots to build independent houses.

The social structure of Gurugram changed drastically. Historically Ahirs, Gujjars, Jats and Punjabis dominated the area but by 2010 residents included migrants from whole of northern belt of India and also expatriates (Mehtani, 2012). These migrants are largely economic migrants who moved to the city due to employment in newly developed BPO/KPO sector. Their cultural and social connection with the city was weak, and they espoused "private services" that the newly developed private townships offered.

The biggest loss to the built environment development philosophy that the period beginning 1990 brought was the loss of public spaces due to the development of private residential complexes. Though the colonies in the HUDA master

Figure 5.1 Kingdom of Dreams

Source: Photograph by Harsh Tripathi, 2017

Figure 5.2 Private residence at DLF City

Source: Photograph by Harsh Tripathi, 2017

Figure 5.3 Essel Towers, Gurugram

Source: Author, 2017

plan did have spaces for community parks, they were seldom used by the local residents who would live around them in multistory apartments with their own private community services. Consumerism brought with it privatisation of public goods – private schools, gated communities, private swimming clubs, private cars, private healthcare, private security, private water system, private sewage collection, private fire system, and private electricity system. While deteriorating pubic services were partly responsible, consumerism hastened the process, as seen in Gurugram, and the government was keen to offload responsibilities for providing public services to the private sector (Doshi, 2016).

5.4.1 *Malls mania*

A new typology that began dotting the landscape of Delhi and its surrounds were "malls", a concept that was imported from the West, for the rising affluent masses of this megacity. Malls became the place where middle- and higher-income classes could shop in a luxury, air-conditioned environment competing successfully with the traditional bazaars and markets of Chandni Chowk or Connaught Place that had survived for hundreds of years. These also replaced the neighbourhood shops that had emerged in different localities post-independence to provide opportunities for settlers after Partition of the country in 1947. The population of Delhi reached 9.42 million in 1991. The first mall in Delhi was built in 1988 as a joint venture between Housing and Urban Development Corporation (a public sector entity) and a private developer, Ansal Group. The mall has unique characteristics, which integrates open public space with indoor shopping area (Figure 5.4).

The population growth in Delhi and its surrounds has been phenomenal. In 2011, the National Capital Region (NCR) has a population of about 40 million. The opportunities for malls are phenomenal, and these became a new layer of

Figure 5.4 Ansal Plaza

Source: Photograph by Harsh Tripathi, 2017

Figure 5.5 DLF City Centre Mall, Gurugram

Source: Author, 2017

building typology in Delhi and its surrounds. Gurugram became one of the fast-growing mall centres post-2000. Two road stretches of the city included almost all 13 malls (Figure 5.5). Today, Delhi and its surrounds have more than 50 malls, and construction of another 40 malls have been planned. The formats of these malls have become far more elaborate, and they not only house shops but also include multiplexes, food court, entertainment areas and hotels. Malls became a hub of family activities. These are the new replacements of open public spaces of the past for meeting, entertaining and shopping.

A similar story can be painted for other fringe townships that have grown around Delhi such as Noida, Faridabad and Ghaziabad.

5.4.2 *Metro rail*

Metro rail is the modern-day large-scale intervention in the built environment of Delhi and its surrounds that has added an important layer to the spatial landscape connecting masses and built places. In some ways metros have also contributed to the development and popularity of malls. Prior to 2002, after which when metro lines became operative, the main mode of transportation in Delhi were private vehicles (two or four wheelers) as the public mode of transportation was poor. The successive governments invested in roads as rings around the city and the development followed these transport rings. Often other developments preceded the development of roads and other infrastructure.

Figure 5.6 Rapid Metro, Gurugram

Source: Author, 2017

Metros completely changed the pattern of development and hastened the growth of fringe cities. While the metro that connects Delhi and its surrounds is financed by the government of India and government of National Capital Territory of Delhi, the Rapid Metro in Gurugram is a public-private partnership project funded and operated by private sector with HUDA as the concessioning authority (Figure 5.6). The initial consortium that had begun building Rapid Metro comprised DLF and Infrastructure Leasing and Finance Company (IL&FS) illustrating the clout that the developers have in determining the built environment in Gurugram. The impact of the metro on the pattern of the built environment development has been profound. High-density developments have emerged along the metro train lines dotted with malls, offices and other commercial developments around stations. As we move away from train stations (or metro line) the development density drops. The density gradient in Delhi and its surrounds, which until 2002 has been in the shape of inverted bell curve with low-density New Delhi and high-density, high-rise development on the fringes of New Delhi, has now been superimposed with another pattern with the high density along the train lines and low density in-between.

5.4.3 *Iconoclast*

The third important element of consumerism that one witnesses is as a result of race among developers to develop iconic buildings and mega townships. These forms have been made possible by compulsory acquisition of large land parcels on the fringes to develop infrastructure projects using public-private partnership models. The Wave City in Ghaziabad and Jaypee Sports City on the Yamuna Expressway, Greater Noida, are some of the examples. Wave City is a large private township development with its own municipal services (Figure 5.7), and Jaypee Sports City promises to develop an iconic city built around 'sports' theme. Formula One car race track is already in place. Between Delhi and Agra on the Yamuna Expressway in locations such as NH24 and Greater Noida, a number of

Figure 5.7 Wave City, Noida

Source: Photograph by Elizabeth Lisinski, 2015

Figure 5.8 High-rise residential development, Noida

Source: Photograph by Elizabeth Lisinski, 2015

Figure 5.9 Wish Town, Noida

Source: Photograph by Elizabeth Lisinski, 2015

Figure 5.10 Advertisements for sale of residential apartments

Source: Photograph by Elizabeth Lisinski, 2015

high-density, high-rise apartment towers have been built which have completely transformed the landscape of rural villages that earlier occupied these areas (Figure 5.8). Developers woo their customers through a promising luxurious lifestyle (Figure 5.9) or through enticing financing options (Figure 5.10).

5.4.4 *(Un)Sustainability*

According to Forest Survey of India conducted by the Ministry of Environment and Forests, in 2015, 20.22 per cent of land area in National Capital Territory of Delhi was covered with trees and forests. The map illustrates that the most of the forest cover is located in New Delhi and along the ridges in the north of Shahjahanabad and the south around Mehrauli – Tughlaqabad. Even though the land area of core New Delhi or Lutyens' Delhi, as it is called, is controlled in development activities and the Ridge Management Board manages ridges, reports document that nearly 40 per cent of the area of ridges was encroached by the colonies that were set up after Partition and some through illegal occupation later (Sinha, 2014). A large part of remaining ridge is managed and protected providing the necessary green cover to Delhi (ibid).

The land to the southwest of Delhi, where Gurugram in Haryana state is located, is however devoid of any major green cover. According to the Forest Survey of India only 8.35 per cent of geographical area of Gurugram is forest. Forest cover in Aravallis ranges plays an important role in maintaining ecological balance of the region by "trapping global warming gases, enhancing water recharge capacity to augment groundwater and reviving water bodies; reducing heat island effect and the impact of extreme weather events" (Roychowdhuri, 2017, p. 40). However, these ranges are under stress due to increasing construction activities.

Prior to the development of the city, it was arid land used primarily for low-productive agriculture production. The development activity that has happened in Gurugram particularly after 2000 has imposed tremendous stress on the natural resources. The consumerist attitude to development prevented the sustainable development practices, urban planning, design and management to gain ground. The United Nations' Sustainable Development Goals, which were adopted by 194 countries in September 2015, lay down guiding principles for urban planning, design and management. These specifically aim at air quality, waste management, accessible and sustainable transport systems particularly through the development of public transport system, upgrade infrastructure, planning towards inclusion, resource efficiency, mitigation and adaptation to climate change and resilience to disasters among others (Roychowdhuri, 2017).

In his report *Gurugram: A Framework for Sustainable Development*, Roychowdhuri (2017) argues that the city faces a number of challenges, which have arisen due to unsustainable development practices that the city has followed over the past few decades. According to an estimate by the Centre for Science and Environment, 137 water bodies have been lost due to development activities. Gurugram gets its water from the River Yamuna. However, this is able to meet only 30 per cent of the water requirement in the city. For rest of the demand for water, households rely on groundwater (ibid). However, the situation with groundwater is precarious. An estimate by the Central Ground Water Board (CGWB) indicates that the water level in Gurugram dropped from 43 m bgl in 2003 to 51 m bgl in 2006. The rate of fall in the water table per year is about 1–3 metres causing a situation classified as "water stress" (Roychowdhuri, 2017). Moreover, the groundwater has become polluted due to "unscientific exploitation of groundwater, agricultural and industrial activities in the vicinity, poor sanitation, inadequate septage management and solid waste disposal" (Roychowdhuri, 2017, p. 16). Though there are restrictions on the extraction of groundwater, unabated withdrawal of water continues. According to an estimate, cited in (Roychowdhuri, 2017), builders extract 50 million gallon of water illegally for construction purposes.

Depending on the source of estimates, Gurugram produces between 129 and 260 MLD of sewage, while the capacity of the city to treat sewage is only 30–50 per cent of the sewage generated (Roychowdhuri, 2017). The failure of public infrastructure has led private developers of apartment complexes and commercial parks to provide waste-treatment services in some of their townships. Roychowdhuri (2017) notes that "once the waste is treated, tanker trucks transport the waste to the Yamuna River or the treated effluents are disposed of into an unlined drain which travels about 8 km to meet the Najafgarh drain in Delhi" (p. 19) causing serious health hazards.

Based on empirical evidence Roychowdhuri (2017) states that "Gurugram is now one of the most polluted cities in the NCR" (p. 21). He identifies that the "challenges are explosive growth in vehicle fleet; massive use of diesel guzzling vehicles and generators; and sharing the toxic air shed of Delhi" (p. 21). The lack of an efficient public transportation system has led to a situation "leading to severe traffic gridlock; an unfolding parking crisis; and insensitive urban design which leads to unsafe roads and pollution lock-in" (Roychowdhuri, 2017, p. 25).

The energy consumption in Gurugram has risen sharply. Power outages are common. Roychowdhuri (2017) reports that in many parts of the city the residents and businesses expect 10–12 hour power outages, especially during summers which has led businesses and residential associations or individuals "to generate most of the electricity, with diesel generators in their backyards, and there are no norms in place for regulation or monitoring of such Generators" (Roychowdhuri, 2017, p. 29).

Rising incomes and changing lifestyle have rendered building design practices and operations to become energy-intensive. Copious construction activity, particularly of the high-rise luxury buildings, has led to an increase in the use of energy-intensive material like glass and steel (Roychowdhuri, 2017). Excessive use of glass in building façade traps heat and increases cooling requirements further (ibid).

The generation and management of solid waste is another serious issue. A large part of solid waste is construction debris. Lack of disposal sites have meant, "these are either dumped in city landfills or in open spaces, water bodies and flood plains" (Roychowdhuri, 2017). There are serious concerns about dumping of construction debris along the roads, water recharge bodies and the Aravallis (ibid).

5.4.5 Mal à propos

> Most persons think that a state in order to be happy ought to be large; but even if they are right, they have no idea of what is a large and what a small state. . . . To

> the size of states there Is a limit, as there Is to other things, plants, animals, implements; for none of these retain their natural power when they are too large or too small, but they either wholly lose their nature, or are spoiled.
>
> —Aristotle, 322 BC (cited in Meadows et al., 1972)

While the interventions in built environment in the past have been ambitious and of mega scale for the time when they were made – such as construction of cities by Sultanate or Mughal rulers or construction of New Delhi – the development post-2000 in Delhi and its surrounds is at a scale and of a style that seems to lack empathy towards the existing context. This seems to respond to a rising income and a lifestyle that is an exemplification of 'consumerism', which implicitly implies that consumption of "more" by an individual is better. This also means privatisation and persuasion of self-interest. The high-rise residential buildings largely invested in by investors for capital gains, use of unsustainable materials which are energy inefficient, the loss of public open spaces and activities, rising social and wealth inequalities propelling the consumerism and privatisation further are all symptoms of modern-day development in surrounds of Delhi.

5.5 Embracing disruptions in time and space: summing up

The book is an inquiry in more than a millennium history of the built environment in Delhi and its surrounds. Using social theoretical approach and lens of institutions the book uncovers the layers of the built environment and the forces – political, economic, social – that shaped the urban form, design and buildings, some surviving in their full galore and others in ruins. However, each of these has a story to tell of ambitions, innovations, beauty, modesty, functionality, utility and whim and fancies of those who conceptualised them. Each new city in Delhi's history wanted to outdo the previous one, however, not in a disconnected fashion but through interconnectedness that provided agility to the city over time and space.

If we join the 'parts', they all come together to form the 'whole', which is modern day-Delhi. The dependencies between the built environment and the humanity seem seamless, as the old buildings have found a new use appropriate for time. Not only the palaces and buildings built by rulers and their descendants but buildings built by many commoners dot the landscape, for example Begum Samru's Palace, the earliest colonial buildings with Greek pillars, in Chandni Chowk (refer to Figure 4.55). It is now used as a branch of Central Bank, although the building is poorly maintained, like many other surrounding structures, together constituting to the dense, mixed-used character of Chandni Chowk. During the early part of the twentieth century, it was also used as a branch of Lloyds Bank, the name of which remains imprinted on the building. The townships and refugee colonies that emerged during the nineteenth and early twentieth centuries to the north of Shahjahanabad (Model Town, Sadar Bazar, Sabji Mandi) form an axis of culture of migrants from west of British India after Partition. A number of markets, such as Khan Market near India Gate in New Delhi or Sarojini Nagar

Market that were community markets to employ migrants affected by Partition, have become one of the most expensive retail properties in Delhi. While the Red Fort ceased to be in use after New Delhi was built as the official seat of government, it has retained its pre-eminent position of power and supremacy as every year India celebrates its Independence Day at the Red Fort. Hauz Khas village in the city of Siri is a bustling hub for artisans with boutique shops of fashion and craft. What is unique about Delhi is the co-existence of modern with ancient with each layer complementing rather than contrasting the other. Metro trains have played their own part in connecting spaces and developing urban form, which is an interesting potpourri resulting in a unique Delhi culture. This reflects the agility of the city. The city has embraced the disruptions that invasions by Turks, Afghans, Persians, Mongols and British in India caused. Delhi became the capital of empires that some of these foreign invaders established in India. They built cities that seem fragmented from the rest in one time but were embraced in larger landscape over time. There is no building or urban region or part of an urban region in Delhi that one could say does not have a modern use and has been abandoned.

The modern-day development, however, poses a challenge as it seems to be fast-paced, lacks context and is investor-driven whose interests are global rather than local. The incessant development that has continued over a millennium has left Delhi and its surrounds in a weak position related to environmental sustainability. In the past, the development offered time to recoup the damage to the environment. However, the modern-day development is not offering that time. The vulnerable environment will be a key disruptive factor for the built environment in Delhi and its surrounds. While Delhi has embraced disruptions in the past, probably by luck, the future would need to be treaded carefully. Delhi is not a singular cell organism and is rather a complex constitution of significantly distinct elements, both physical and non-physical (social, religious, cultural) bound together in a functional way. Whether these variations, as they exist today, are to be celebrated or should these be harmonised is open for debate.

We would close the book by asking questions once raised by John Stuart Mill:

> Towards what ultimate point is society tending by its industrial progress? When the progress ceases, in what condition are we to expect that it will leave mankind?
>
> —John Stuart Mill, 1857 (cited in Meadows et al., 1972)

Bibliography

Breton, A., Seaver, R. & Lane, H.R. (1969). *Manifesto of Surrealism*. Ann Arbor Paperbacks.

Doshi, V. (2016, July 4). What Life Is Like in the Indian City Built by Private Companies. *The Guardian*.

Meadows, D. H., Meadows, D.L., Randers, J. & Behrens III, W.W. (1972). *The Limits to Growth*. New York, NY: Universe Books.

Mehtani, P. (2012). *Growth, Development and Sustainability of Cities: A Case Study of Gurgaon*. New Delhi: University of Delhi, Department of Geography.

Roychowdhuri, A. (2017). *Gurugram: A Framework for Sustainable Development*. New Delhi: Centre for Science and Environment.
Sen, A. (1979). Utilitarianism and Welfarism. *The Journal of Philosophy*, 463–489.
Sinha, G. (2014). *An Introduction to Delhi Ridge*. New Delhi: Department of Forests and Wildlife, Government of NCT Delhi.
Wright, E. O. & Rogers, J. (2015). *American Society: How It Really Works*. New York, NY: W.W.Norton & Company Inc.

Index

Printed in the United States
by Baker & Taylor Publisher Services